Studies in Systems, Decision and Control

Volume 369

D1826900

Series Editor

Janusz Kacprzyk, Systems Research Institute, Polish Academy of Sciences, Warsaw, Poland

The series "Studies in Systems, Decision and Control" (SSDC) covers both new developments and advances, as well as the state of the art, in the various areas of broadly perceived systems, decision making and control–quickly, up to date and with a high quality. The intent is to cover the theory, applications, and perspectives on the state of the art and future developments relevant to systems, decision making, control, complex processes and related areas, as embedded in the fields of engineering, computer science, physics, economics, social and life sciences, as well as the paradigms and methodologies behind them. The series contains monographs, textbooks, lecture notes and edited volumes in systems, decision making and control spanning the areas of Cyber-Physical Systems, Autonomous Systems, Sensor Networks, Control Systems, Energy Systems, Automotive Systems, Biological Systems, Vehicular Networking and Connected Vehicles, Aerospace Systems, Automation, Manufacturing, Smart Grids, Nonlinear Systems, Power Systems, Robotics, Social Systems, Economic Systems and other. Of particular value to both the contributors and the readership are the short publication timeframe and the worldwide distribution and exposure which enable both a wide and rapid dissemination of research output.

Indexed by SCOPUS, DBLP, WTI Frankfurt eG, zbMATH, SCImago.

All books published in the series are submitted for consideration in Web of Science.

More information about this series at http://www.springer.com/series/13304

Aboul Ella Hassanien · Ashraf Darwish ·
Benji Gyampoh · Alaa Tharwat Abdel-Monaim ·
Ahmed M. Anter
Editors

The Global Environmental Effects During and Beyond COVID-19

Intelligent Computing Solutions

 Springer

Editors
Aboul Ella Hassanien
Faculty of Computers and Artificial
Intelligence
Cairo University and Scientific Research
Group in Egypt (SRGE)
Giza, Egypt

Benji Gyampoh
Department of Fisheries and Watershed
Management
Kwame Nkrumah University of Science
and Technology
Kumasi, Ghana

Ahmed M. Anter
Faculty of Computers and Artificial
Intelligence
Beni-Sweif University
Bani Sweif, Egypt

Ashraf Darwish
Faculty of Science
Helwan University and Scientific Research
Group in Egypt (SRGE)
Helwan, Egypt

Alaa Tharwat Abdel-Monaim
Faculty of Computer Science
and Engineering
Frankfurt University of Applied Sciences
Frankfurt am Main, Germany

ISSN 2198-4182   ISSN 2198-4190 (electronic)
Studies in Systems, Decision and Control
ISBN 978-3-030-72935-6   ISBN 978-3-030-72933-2 (eBook)
https://doi.org/10.1007/978-3-030-72933-2

This Springer imprint is published by the registered company Springer Nature Switzerland AG
The registered company address is: Gewerbestrasse 11, 6330 Cham, Switzerland

Preface

Recent advances in technology will play a significant role in tackling the Coronavirus (COVID-19) pandemic in order to understand the virus and accelerating medical research on treatments and drugs, diagnose and detect this disease. Morever, predicting its evolution assists in preventing or slowing the virus'spread through surveillance and contact tracing, respond to the health crisis through personalized information and learning, and finally monitor the virus recovery and improving early warning tools.

While the pros and cons of COVID-19 on the environment are evident in the literature, the environment or climate has also significantly impacted COVID-19 transmission and deaths. COVID-19 has become a global epidemic. Its relationship to environmental factors is an issue that has caught the attention of scientists and governments. Intelligent computing plays a crucial role in better understanding and addressing the COVID-19 crisis and being used as a tool to support the fight against the viral pandemic that has affected the entire world since the beginning of 2020. This editing book aims to collect chapters on using intelligent computing technologies to handle environment-related problems of COVID-19.

This book aims to present the role of recent technologies such as Artificial Intelligence and the Internet of Things to fight against COVID-19 to make it easy for researchers and academics to investigate new techniques that will help the environment and healthcare sector workers and reduce the spread of the COVID-19 pandemic. This book also presents the problems and challenges and some future research points from the recent technologies point of view that can help the environment and healthcare sectors reduce COVID-19.

Finally, editors of this book would like to encourage the researchers, readers, and specialists to explore this book's main topic to provide a sustainable environment and healthcare sector and create their implementations and obtain recent results to tackle COVID-19 global pandemic.

Cairo, Egypt
Helwan, Egypt
May 2021

Aboul Ella Hassanien
Ashraf Darwish

Contents

The Role of Internet of Medical c in Monitoring the Environmental Impact of COVID-19 1
Doaa Mohey El-Din, Aboul Ella Hassanien, and Ashraf Darwish

COVID-19 and Water Resources Nexus: Potential Routes for Virus Spread and Management Using Artificial Intelligence Techniques 19
Hend S. Abu Salem, Mahmoud Y. Shams, Aboul Ella Hassanien, and Ahmed M. Nosair

Environmental Challenges and the Impact of COVID-19 on Healthcare Sector: The Adoption of Intelligent Solutions 41
Yasmine S. Moemen and Ibrahim El-Tantawy El-Sayed

COVID-19 Health Waste Management in Taiwan 55
Kai-Chun Chu and Kuo-Chi Chang

Artificial Intelligence for Sustainable Waste Management and Control During and Post COVID-19 Crisis: Critical Challenges ... 81
Walid Hamdy, Ashraf Darwish, and Aboul Ella Hassanien

Carbon Monoxide Air Pollution Monitoring Approach in Africa During COVID-19 Pandemic 93
Reham Gharbia and Aboul Ella Hassanien

Applications of Deep Learning in Predicting Natural Disasters Concurrent with the COVID-19 Pandemic: Short Review and Recommendations 105
Dalia Ezzat, Sara Abdelghafar, and Aboul Ella Hassanien

Sustainable Climate Change Policies Driven by Global CO_2 Reduction During COVID-19 121
Haytham H. Elmousalami

**The Correlation Between Human Lifestyle, Weather, Smart
Technologies and Covid-19 Pandemic** 137
Yasmine S. Moemen, Ibrahim El-Tantawy El-Sayed, Ashraf Darwish,
and Aboul Ella Hassanien

Exploring the Impacts of COVID-19 on Oil and Electricity Industry ... 149
Atrab A. Abd El-Aziz, Nour Eldeen M. Khalifa, and Aboul Ella Hassanien

**COVID-19 Outbreak and Its Effect on Global Environment
Sustainable System: Recommendation and Future Challenges** 163
Amira S. Mahmoud, Mahmoud Y. Shams, and Aboul Ella Hassanien

The Role of Internet of Medical c in Monitoring the Environmental Impact of COVID-19

Doaa Mohey El-Din, Aboul Ella Hassanien, and Ashraf Darwish

Abstract COVID-19 (Corona Virus disease 2019) is the highest spreading virus around the world. October 2020, more than 30 Million around the world are infected by this virus. A rising number of infected people number with disease's reasons the incapability challenge for totally caring in healthcare centers and afflict many physicians and healthcare members inside the hospitals. So, healthcare does not reach all infected patients. So, there is a deep need to guarantee suitable healthcare for patients remotely and save their lives. Monitoring patients in healthcare centers is considered the most significant medicinal situation. This process is an infectious cause of thousands of healthcare workers, so there is a high requirement for remote monitoring. On another side, less critical patient's side, this system permits observers, doctors, and nurses to monitor patients and acquire healthcare from patient's homes to keep places for the crucial situations in hospitals. To save the Individual's lives by improving health services from infection. The smart application monitors the infected people who rely on several devices to register their patients' properties every second. This paper presents a review study of the importance of Internet-of-things in smart health applications. It also shows the beneficial roles in providing medical issues for infected patients remotely based on monitoring electronic Quarantine (E-Quarantine) remotely. It benefits sensory big data interpretation to support physicians in patients' diagnosis via the Internet.

Keywords Internet-of-Things, sensors · Artificial intelligence · Big data · Quarantine · COVID-19 · Smart health

D. M. El-Din (✉) · A. E. Hassanien
Faculty of Computers and Artificial Intelligence, Cairo University, Cairo, Egypt
e-mail: d.mohey@alumni.fci-cu.edu.eg

A. Darwish
Faculty of Science, Helwan University, Helwan, Egypt
e-mail: ashraf.darwish.eg@ieee.org

D. M. El-Din · A. E. Hassanien · A. Darwish
Scientific Research Group in Egypt (SRGE), Cairo, Egypt

© The Author(s), under exclusive license to Springer Nature Switzerland AG 2021
A. E. Hassanien et al. (eds.), *The Global Environmental Effects During and Beyond COVID-19*, Studies in Systems, Decision and Control 369,
https://doi.org/10.1007/978-3-030-72933-2_1

1 Introduction

Recently, Coronavirus (COVID 19) has a tremendous spreading infection worldwide that reaches more than 40,534,313 million around the world in October 2020 [1]. The healthcare is not suitable enough to guarantee all infected patients. So, the infected patients achieve a large number of people. In October 2020, the deaths caused by Corona Virus reached 1,121,444 around the world [1, 2]. The number of infection pervasion people and deaths numbers are rising daily.

People who usually have a high impact on patients are healthcare members whether doctors or nurses. The highest infection of COVID-19 statistics is shown in healthcare members such as doctors and nurses [3]. Thousands die from the COVID-19 infection, whether mild or mixed patients, in healthcare centers. So, there is a deep requirement to proceed the remote healthcare. Monitoring healthcare is considered a remote service for multiple patients. It also needs many sensors to save each case's parameter in real-time to enhance healthcare services rapidly and make decisions remotely [4]. Figure 1 shows the cumulative statistics of infection people from COVID-19, deaths from infection, and treatment.

There is a shortage of healthcare workers and equipment worldwide, so there is a deep need for using internet-of-things technologies and smart sensors to simulate the monitoring systems remotely [5]. The recent smart application aims to support healthcare services such as observing after surgical actions, observing kids remotely, and monitoring chronic diseases. The construction of any smart application depends on the connection of sensors and smart devices via the Internet. That uses the power of Internet-of-things technologies and artificial intelligence from big sensory data [6, 7]. These big data have a vital role in interpreting data and fusing data. Internet-of-Things is defined by a well-defined scheme that depends on connecting sensors via the Internet. It is based on machines sensors possessing and computing tactics, and machine learning. The IoT environments are implemented based on minimizing healthcare costs and enhancing the medication result of the stomachic patient. Figure 2 depicts the main benefits of adopting IoT in the COVID-19 pandemic.

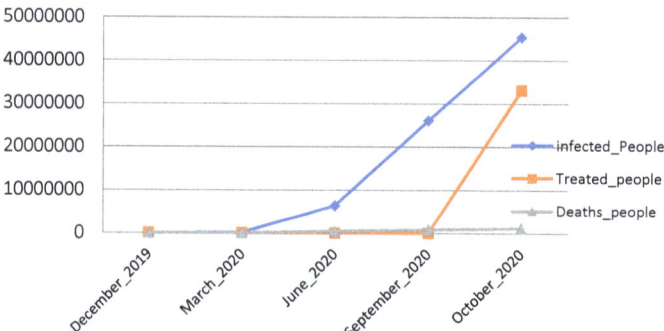

Fig. 1 The ratio of increasing infected patients, deaths, and treated people from COVID-19

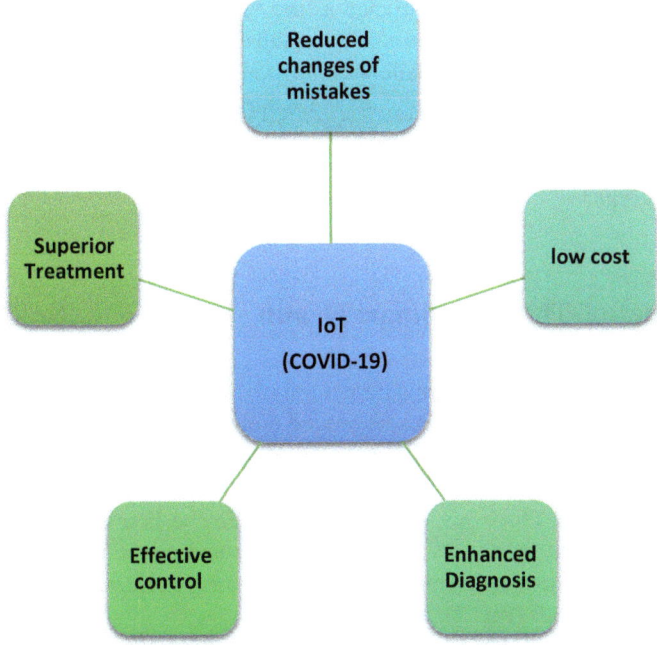

Fig. 2 The main benefits of adopting IoT in the COVID-19 pandemic

Artificial intelligence is considered the ability of a digital computer or computer-controlled robot to perform tasks commonly associated with intelligent beings. The extracted sensory data reaches thousands or million records that includes time mostly per-second and the data. These big data play a vital role in managing and controlling remotely. Previous research aims to monitor the patient in different diseases, such as diets, surgical processes, and debates. They allow doctors to monitor multiple patients simultaneously. That provides facilities for healthcare applications to be very adaptable and careful. The utilized sensors are various applications such as wearable smart devices or mobile sensors [5]. These applications need to explain the extracted sensory data from multiple sensors to achieve the essential target.

This chapter introduces a survey of the importance of using internet-of-things in building smart health applications. E-Quarantine applications enable physicians to track diseases and sudden health problems used to save lives, cost, and time. These applications become very important for thousands of remote villages that require a long time to go to the right doctor and high transportation cost, with an increasing number of infected people with COVID-19. Monitoring infected people requires a huge number of hospitals, nurses, equipment's, and doctors. So, healthcare for COVID-19 patient becomes very hard. It becomes serious about keeping thousands of lives from infection or death. It relies on collecting ever data from various sensors to determine the disease development of the health condition.

The rest of the chapter is constructed as the following. Section 2 discusses the importance of internet-of-Things in a smart health environment. Section 3 shows the Internet-of-Things benefits and challenges. Section 4 examines a comparative study between previous smart health applications for various objectives; Sect. 5 discusses the Role of Internet-of-Things for COVID-19. Section 6 presents open research challenges and future directions. Finally, Sect. 7 concludes the outline's conclusion and future works.

2 Internet-of-Things in Smart Health

Internet-of-Things means simulating the natural environment based on physical thousands or millions or billions of sensors around the world that are connected via the Internet. It uses for improving decision making remotely [4, 5]. Internet-of-things is very powerful for simulating Smart IoT environments [6, 7]. Any IoT Environment is built on four main processes, as shown in Fig. 2. The Internet of Things (IoT) refers to review analysis of previous related works and methodologies, protocols, and applications in this newly emerging field. This survey research presents the IoT taxonomy and methods (Fig. 3).

According to World Health Organization (WHO), Sustainable Development in healthcare worldwide is achieved to the highest investment in 2030 [8]. Smart Health

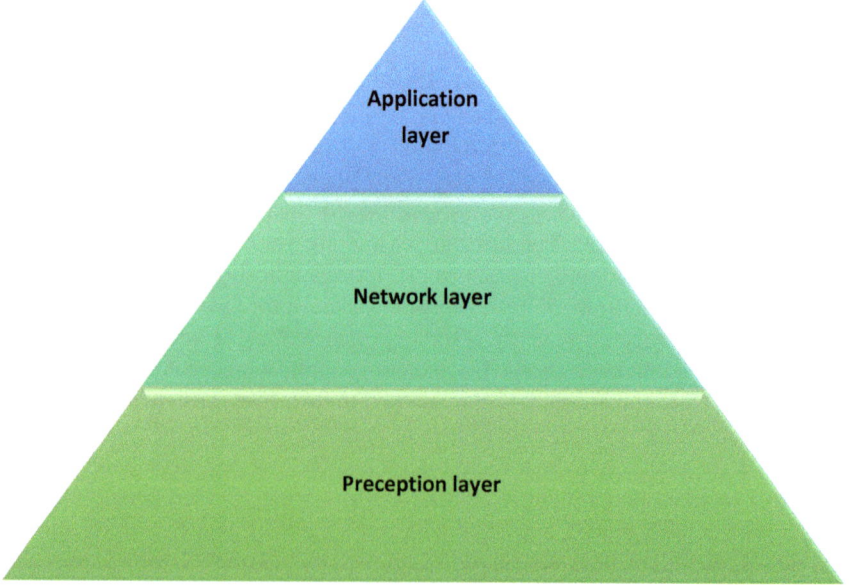

Fig. 3 The Internet-of-Things architecture

Fig. 4 The main three dimensions for construct smart health applications

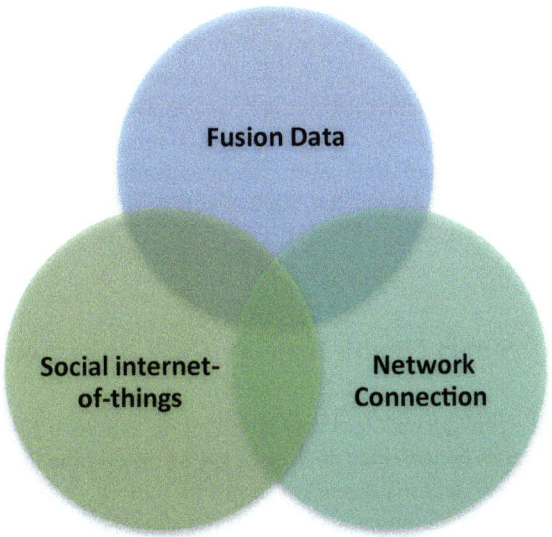

is constructed based on smart devices or sensors to simulate real healthcare applications with one or multiple users. Three main dimensions are designed for reaching the main objective like the following (as shown in Fig. 4), fusion sensory data, network, and Social Internet-of-Things. The hardest step of any smart health application is a fusion. It includes preprocessing, cleaning data, and combining sensory data with multiple targets to unify and understand the main objective. The network requires connecting availability all the time. It requires securing data and the network from any hacking or outliers. Social Internet-of-Things (SIoT) is a hot area of research in smart health. SIoT is defined by multiple users that have various permissions for the same application. For instance, monitoring chronic diseases patients' application can be observed based on multiple users (physicians and nurses).

Prior motivations aim to save cost and performance time. They improve decision-making based on the benefit of connecting smart devices via Internet. The essential target of smart IoT application is checking the observing their sickness pursue in the current time. Previously, finding to build any smart health application needs understanding almost conditions and some expert knowledge to keep automatically and determine vital readings or anomalies for each patient. That needs supervised learning to help any current situations and determine challenges. Visualizing data is used to show metadata about various patients, which is very powerful in saving time. It also is used for keeping lives simultaneously.

Prior studies introduce multiple smart health contributions for building a convenient application for monitoring remote patients based on each healthcare case. Researchers in Galletta et al. [9] construct a graphical system for smart health used to visualize patient data for remotely physicians. The main disadvantages are noisy data and redundant properties. Researchers in Radovic et al. [10] introduces a new smart health application to peak the accuracy for observing patients after operations or surgeries. This application needs healthcare experts and large analysis from doctors to fill the gap vision of various cases. Researchers in Kuang and Davison [11] enhances the monitoring of patients remotely with an accuracy of 9%. They face many disadvantages in reliability and integrity. Researchers in Khoie et al. [12–14] construct smart medical applications for healthcare centers to help observe patients remotely. However, this issue is still having many contributions for improving the accuracy results.

3 Smart Health Benefits and Limitations

Smart Health becomes very powerful for monitoring patients and tracking diseases. It depends on wearable devices based on the chronic illness. The smart health system's benefits and limitations are shown in Table 1.

These challenges and benefits are shown in various contexts and environments as the following.

(a) **Smart health Benefits**

It includes several benefits from using IoT, which are shown in Fig. 3. The improvements through using Internet-of-Things are developed that support managing remotely for any domain Favaloro [15]. The essential benefits of using Internet-of-Things. It is highly effective management that can control automatically in real-time.

Table 1 A comparison between Internet-of-Things challenges and benefits in constructing multiple smart environments based on five dimensions

Comparison dimension	IoT challenges	IoT benefits
Data size	The hardness of big data and Interoperability process from sensors. (variety, velocity, volume, and variety)	Effective fusion for reaching the main objective of smart application
Cost issue	It has some challenges in the first time to simulate any smart sensors due to choose the real smart environments	It reduces cost of simulating application such as the minimizing cost of real visits to doctors or hospitals
Security issue	The difficulty of hacks and attacks on the network that are connected all the time	Multiple users controls in real-time (Social Internet-of-Things)
Business model contexts	It is requires awareness about each business model, features, and conditions	It requires optimizing performance for any business model
Noisy of data	The challenges of classification types of outliers	Detect outliers and classify errors

Using IoT has cost reduction and reveals management controls for multiple users. It reduces errors and detects outliers from extracted sensory data. It improves monitoring IoT that monitors patients and tracks diseases remotely. It improves a predictive analysis for collecting large data sizes, IoT's latest technologies. This delicate information will be utilized to enhance current operations and services. It builds a speed response that relies on data due to following the applications concurrently and remotely. They optimize the maintenance interventions that use for improving the strategies of any market [16]. It reduces human errors for the complementary methodologies as artificial intelligence (AI).

(b) **Smart health Limitations**

Smart health's main limitations are explaining, fusing, and visualizing big data extracted from many smart sensors. It enhances the right decisions for various healthcare environments. The data are merged from smart devices to monitor patients remotely at the Quarantine's homes. These data are used by gathering the statistical methods for the decisions of the healthcare application. It also introduces the challenges of IoT environments.

4 A Comparison Between Smart Health Applications

Smart Health is a hot research field of research and industrial applications which include a connection among various smart sensors to each patient. It monitors remotely for the patient in several data types, for example, video, audio, or text. The essential challenge controls data analytics and visualizing data. Smart Health or telehealth refers to simulate real medical smart environments remotely. There are different objectives of various smart health applications, for example, smart hospital, smart pharmacy, Electronic-quarantine, tracking diseases, or monitoring patients, as shown in Fig. 4. These depend on many users such as doctors, patients, pharmacies, labs, images, hospitals, and ambulance. It depends on multiple smart devices or sensors for implementing smart health applications such as Thermometer, blood pressure, blood glucose, ECG, or smart wearable for various diseases (T-shirt, watch) as shown in Fig. 5.

The recent motivations in smart health are classified into three types: web applications, smart wearable, mobile applications—the following study of review analysis about smart health systems for various purposes. Remote patient monitoring (RPM) is defined by an evaluation analysis of a patient's that relies on context target and dimensions of health metrics such as heart rate, blood pressure, body temperature, have chronic diseases or not, and medication history. These applications have three smart health types, web application, mobile application, and wearable applications, as the following in Fig. 6.

There are prior survey studies about smart health applications for enhancing the healthcare remotely and supportive data for patients [17, 18] as shown in Table 2.

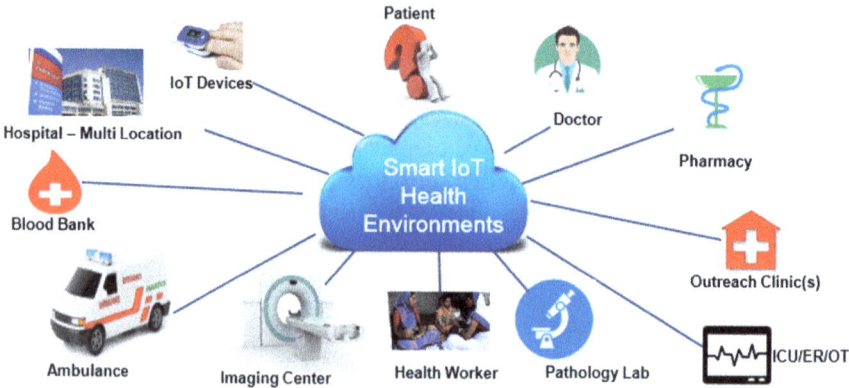

Fig. 5 IoT for smart health environments

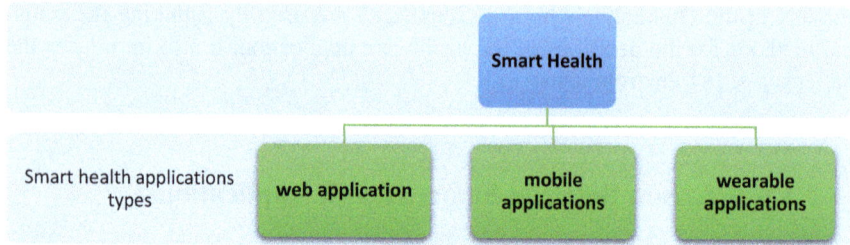

Fig. 6 Smart health applications are classified based on three types

Table 2 A comparison study of multiple smart health applications based on mobile applications

References	Objective	Methodology	Pros	Cons
[19]	Monitoring mobile application	Using noise reduction, segmentation, feature extraction, anomaly detection, and visualization	It presents a powerful of mobile application High flexibility	Complexity
[20]	Supportive data for patients based on mobile application	Machine learning algorithms	It can search the available and suitable hospitals automatically	Complex algorithm
[21]	Monitoring mobile application	Body Mass Index(BMI)	Predication generation for patients. It is a quick and effective way	
[22]	Monitoring mobile application	Greedy strategy, machine learning, and naïve tag	Improves detecting diagnosis of several diseases based on health conditions	Complex implementing

Researchers in Mora et al. [19] proposed a smart health framework for monitoring patients based on Mobile android environments. It utilizes biomedical data's in BSN. The experiment can extract data from wearable devices, smart biosensors, and the cloud for storage and mobile environment. Researchers in Imteaj and Muhammad [20] designed a mobile application for patients, searching for the most suitable and nearest hospitals by patients. It also introduced help by first aid information prior. It becomes very powerful to get data without the Internet due to depending on the server. Researchers in Yi and Saniie [21] presented a mobile Healthcare application for monitoring patients designed based on communicating the various biosensors. It generated prediction signs of the patients for enhancing the integrity and analysis. Researchers in Jeong et al. [22] presented a smart healthcare system for monitoring patients based on using the internet-of-things. It is used for detecting diagnoses for various diseases based on health conditions.

Another type of smart health application is web applications or web portals [23–25]. Several infected patients require experience in deal with mildly infected with moderate symptoms or similar with cold. It is essential to monitor infected people based on the COVID-19 disease nature and their health status changes. Research in Lu and Zhang [26] presents the importance of E-health and the main challenges of security, integrity, and preprocessing data on the web sites applications. Researchers in Mwaffag et al. [27] constructed a smart framework for COVID-19 patient's information that uses machine learning techniques for making prediction analysis of the disease. They use the real dataset for infected people and use Support Vector Machine, Neural Network, Naïve Bayes, K-Nearest Neighbor (K-NN), Decision Table technique, Decision Stump technique, OneR, and ZeroR. The accuracy results are reached more than 90%. Researchers in Alshraideh et al. [28] presented a detecting technique using patient demographic data as inputs, along with based on ECG signal features extracted automatically through signal processing techniques. The accuracy results reach 98.29% for the Cross-validation. It is also combined a web-based system that can be utilized at any time by patients to check their heart health. Researchers in Talakala and Hari Krishna [29] presented a monitor system for patients based on multiple dimensions: ECG, Electromyogram (EMG), pulse, blood glucose, and body temperature. It uses for making web server systems and that is compatible to be an android platform. That uses for remote managing and checking framework without GPS, naturally acquire the situation without condition. Researchers in Raj et al. [30] constructed a smart health application based on using IOT based EMG observing gadget. It provides the flag of EMG parameter in smart health applications (Table 3).

Wearable devices are smart devices or sensors that use for improving decision making remotely. They use artificial intelligence and internet-of-things to monitor patients and support their with suitable data such as smart Watch and AirPod headphones, and Wear OS watches from Google [31, 32]. The wearable technology supports increasing the digital transformation for preprocessing data in real-time. These devices allow users to analyze, get experience from users' skills, reduce process costs, and save substantial time [33]. Wearables rely on a wide spectrum of enterprise applications based on wearable's-as-a-service (WaaS). Researchers in

Table 3 A comparison study of multiple smart health web applications

References	Objective	Methodology	Pros	Cons
[27]	Monitor healthcare system	Support Vector Machine, Neural Network, Naïve Bayes, K-Nearest Neighbor (K-NN), Decision Table technique, Decision Stump technique, OneR, and ZeroR	Getting infected patients data and prediction analysis of each case The accuracies reach more than 90%	Complex
[28]	Monitor healthcare system	Machine learning	High accuracy 98%	
[29]	Monitor healthcare system	A Mobile Health Monitoring System (IMHMS)	Powerful for multiple dimensions for infected patients High performance	Upcoming serious diseases
[30]	Monitor healthcare system	Using the repetitive manual lifting	Observing IoT dimensions with medical parameters	Requires training with multiple datasets

Teng et al. [34] introduced a smart health application for health based on wearable energy harvesting. It uses for the measurement of wearable medical sensors. Researchers in Omoogun et al. [35] presented a probable key in forming a wearable wireless monitoring device for evaluating the body temperature, patient's pulse rate and working together mutually over cellular network. Researchers in Liang and Yuan [36] introduced an Electronic Patch that contains new optical biomedical devices, microelectronics, and many smart sensors for constructing wearable health monitoring system. Researchers in Haahr et al. [37] presented a health monitoring system that uses smart physiological devices and transmission sections (Table 4).

That concludes the main problem in prior motivations with variant smart health application types. The most used techniques are statistical methods, computational intelligence techniques, and knowledge-driven techniques as shown in Fig. 7. The statistical methodologies include Hidden Markov Models (HMM), Bayesian Network, Naïve base, conditional random fields, multiclass logistic regression. Computational intelligence techniques include neural network algorithms, support vector machines, decision tree techniques, and clustering. Knowledge-driven techniques include rule-based, fuzzy logic, and ontologies.

Table 4 A comparison study of multiple smart health applications based on wearable devices

References	Objective	Methodology	Pros	Cons
[34]	Introduce p-health system based on wearable energy from smart devices	A review analysis about various smart health applications based on wearable energy	Evaluation energy	Lack of details about methodologies
[35]	Measuring probability from devices for monitoring patients	Probability techniques	Evaluating probability analysis	Requires enhancing results
[36]	Present smart monitoring system	New optical biomedical devices, microelectronics and many smart sensors	Improving decision making	Complex
[37]	Present smart monitoring system	Uses for physiological smart devices, transmission sections	Improving decision making. Electronic Patch that contains new optical biomedical devices	Lack of data

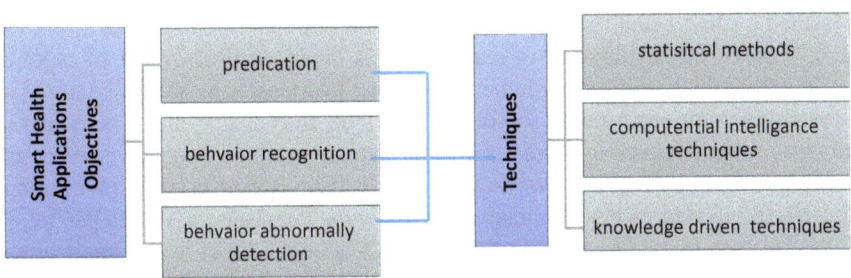

Fig. 7 The main objectives of constructing all smart health applications with used techniques types

5   The Role of Internet-of-Things for Monitoring COVID-19

A coronavirus is considered an evolution of one of the respiratory tract infectious diseases group. The ranges of respiratory are similar between cold and pneumonia. Recently, most infected humans have become a mild disease. These contain the SARS-COV coronavirus. Coronavirus (SARS-COV) was first defined in China in the year 2003. The world countries have been facing the problem of COVID-19 since the end of 2019. During Quarantine, it is useful for a convenient monitoring system, as shown in Fig. 8. All high-risk patients are crossed easily utilizing the internet-based network [38, 39]. The constructing of smart health applications that use for monitoring and COVID-19 diagnosis is mention in Fig. 6.

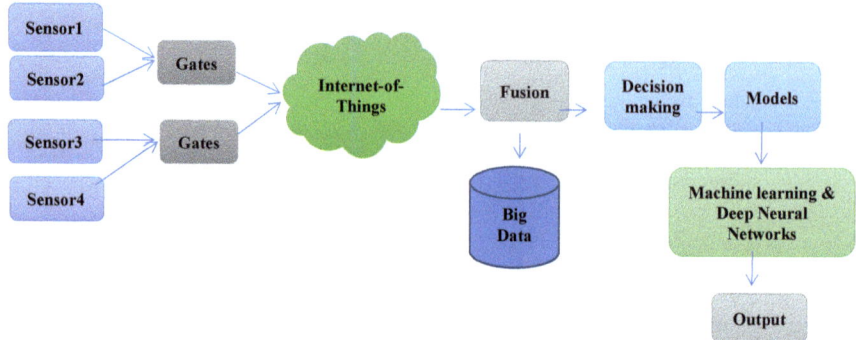

Fig. 8 The Essential Structure Monitoring infected patients with COVID-19

IoT uses for constructing the monitoring systems that use for the infected patients. Previous research presents smart healthcare applications and benefits from Internet-of-Things and artificial intelligence for simulating real smart health applications with real doctors' experiences. That provides filling a gap of medical workers with smart sensors and IoT devices.

Several internet-of-things applications can serve COVID-19 diagnosis, monitoring, and prediction analysis. Researchers in World Health Orgnization [40] presented a smart healthcare application based on Internet-of-Things that monitors patients based on health signs in real-time. It is based on five sensors to capture the data from the hospital environment, including heartbeat sensors, body temperature sensors, room temperature sensors, CO devices, and CO2 devices. Smart health monitoring application is used for tracking the patients remotely based on connecting multiple sensors and smart devices. Recent researches are focused on monitoring COVID-19 infected patients remotely and predicting the progress of each patient.

Researchers in El-Din [41] presents an E-Quarantine smart health application that observes the disease following and predicts the contingency cases around 24 h by 98.7% accuracy results. It relies on five features, blood PH level, the rate of heart pressure, the patient's temperature, and the rate of patients respiratory. They presented a hybrid fusion that depended on a hybrid feature fusion and decision fusion that improved 98.7% accuracy.

Appling a Dempster-Shafer becomes a high benefit based on using sequenced data. These results are better than images or videos. The fusion technique is exercised based on sequenced data for patients and their respiratory sounds. For future work, the proposed healthcare application needs higher flexibility to be adaptive with multiple data types to enhance each patient's results. Researchers in Noun and Hemanvtha [42] aim at saving thousands of lives from infection or death. Recent research introduces an application of E- Quarantine for COVID-19 patients to get healthcare remotely based on five dimensions, body analysis (heart pulse, respiratory rate, body temperature), blood (blood pressure, and blood PH level) with images and text. It can detect outliers of patient's readings in real-time. It experiments on two datasets, the Cardiovascular Disease dataset and Respiratory Sound Database, that depend on fusing two data types, text and audio. They use a type of artificial neural network technique, long-short-term-memory (LSTM) and Dempster-Shafer technique. The results reach 98.7% based on the supervised prior data.

Researchers in Dan et al. [43] introduced a scalable and powerful IoT framework based on using the cloud-assisted for tracking COVID-19 diseases for multiple users as doctors and nurses. It depends on medical parameters as blood glucose levels, body temperature, and oxygen saturation levels.

Researchers in Kumar et al. [44] presented a review analysis about monitoring COVID-19 patients. There is an enhancement of monitoring patients based on questionnaires. That uses for minimizing the Covid-19 spreading using Internet-of-things methodologies.

Researchers in Swati and Chandana [45] present detecting COVID-19 diagnosis and monitoring diseases simultaneously. It introduced an Internet of Things (IoTs) application in real-time that uses symptom data to define the coronaviruses' infection disease. It monitors a medication for recovering the virus and to identify the virus's nature. It is constructed based on the following five essential components: Symptom, Uploading wearable sensors, Quarantine Center, Data Analysis Center, Health Physicians, and Cloud Infrastructure. Researchers in Swati and Chandana [45] discussed the importance of E-quarantine at home or quarantine stations and followed the results of COVID-19. It classifies IoT's main objectives for smart health into seven classes as the following, concurrent-time tracking, remote observing patients, quick diagnosis, contract tracing, and clustering, screening and controlling, minimizing workload of the medical industry, and preventing and management.

Machine learning techniques and deep learning techniques are used for identifying the automatic diagnosis of coronavirus cases in real-time [46, 47]. They identify COVID-19 and monitor patients based on the type of symptom, mild, mixed, infected. The eight algorithms of machine learning are including the Support Vector Machine (SVM), Neural Network (NN), Naïve Bayes (NB), K-Nearest Neighbor (KNN), Decision Table (DT), Decision Stump, OneR, and ZeroR. Other motivations predict infected COVID-19 and the increasing infection between people and animals. Multiple hybrid techniques are used for constructing smart health applications (Table 5).

Previous motivations present multiple benefits for simulating monitoring COVID-19 disease, as shown in Fig. 9. These benefits are demonstrated in monitoring patients in multiple users as defined by social Internet-of-things. IoT uses for recommending systems, recognizing disease as X-Ray, detecting anomaly, and clustering disease levels.

Table 5 A comparative examination between multiple studies in IoT applications for COVID-19

	Objectives	Methodology	Results
[41]	Creating an E-Quarantine smart health application	Dempster-Shafer with neural network	97.8% accuracy results
[42]	It targets monitoring the patient's based on health signs in the real-time It also aims at tracking the patients remotely based on connecting multiple sensors and smart devices It aims at keeping thousands of lives from the infection or death	Machine learning	87.1% accuracy
[43]	Creating IoT framework for tracking COVID-19 diseases for multiple users as doctors and nurses	SVM, KNN, Naïve base classifier	80, 74 and 79%

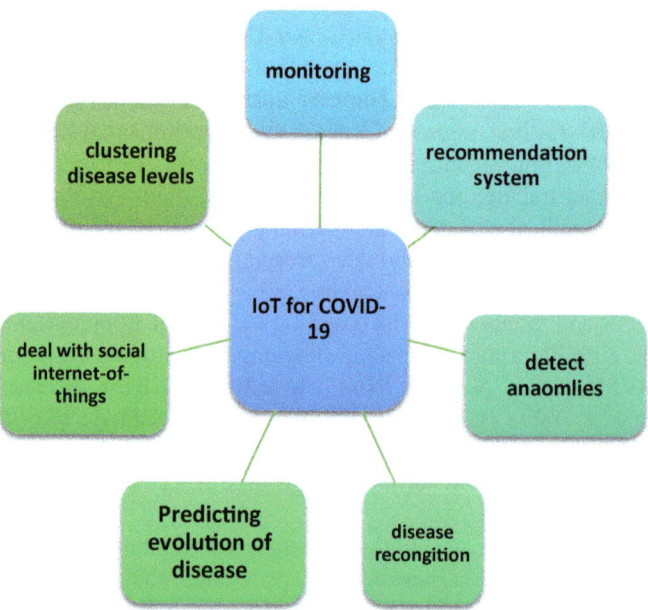

Fig. 9 The benefits of using Internet-of-Things in COVID-19 smart health applications

6 Open Research Problems, Challenges, and Future Directions

From previous motivations, the management of the COVID-19 international pandemic faces many challenges. The Internet-of-Things provides several solutions techniques for constructing healthcare applications to monitor infected patients and save their lives remotely and concurrently. The privacy and security, and fusion of big sensory data are essential to simulating the smart health applications for COVID-19. In the next, the main problems and challenges have been proposed.

- The network Security. The network is connected from various sources with variant protocols. It faces challenges in attacks and hacks that require many motivations for securing the network. Other challenges shown in the construction networks are low power sensors, energy, and the reliability communications challenge.
- Patients' privacy is a single view of patient health. This is the main interest for a future research challenge.
- The sensory data fusion. The communicated networks cause extracting big heterogeneous data that is used to interpret the unification objective of smart health. It uses data from various data sources with multiple data types.
- Data aggregation is considered an important issue in this environment.
- Merging data is based on the constructing network among smart sensors or devices that implicit protocols challenge. The telecommunication protocols have attacks concerning interconnect heterogeneous data that are extracted from the multimodal network.
- The implementation of low power sensors, energy efficiency, and trust telecommunication are the future benefits of this work.
- The IoT wearable smart sensors depend on smart fabrics.
- For mental situation evaluation, emotion-aware Capabilities can be gathered in IoT, which can observe the patient's mental health during the COVID-19 pandemic. They support many personalized therapies.
- Lack of data centers. The governments should provide information and data exchange methods for more investigation and study.
- The main challenge of lack of manpower workers, the lack of healthcare staff, and several experts to deal with the COVID-19 pandemic.
- Lack of infrastructure. Non-adaptable storage field, capacity conditions, poor infrastructure for IoT networks to deal with COVID-19 pandemic.
- Lack of medical facilities. There is still a lack of wearing kits, sanitization capabilities, and disinfection of goods as COVID-19 is pervasion with people contact. It is vital to healthcare capabilities such as the face masks and overall sanitization in organizations and works.
- Lack of access through additional distance for storage concurrently. This challenge is higher about government policy, security, and safety of fighting COVID-19.
- The ethical things in using IoT technology with COVID-19 lie as an open problem.

7 Conclusion and Future Work

This chapter introduces a review analysis about the importance of Internet-of-things in constructing smart health applications. It presents a comparison between multiple smart health systems and classifies them into three types, web applications, mobile applications, and wearable applications. It presents a review study of the importance of monitoring systems for COVID-19 infected patients based on creating E-Quarantine systems remotely. They use for minimizing the infection and protecting hospitals and tool for high-risk patients only. For constructing smart health applications, future motivations are required when using optimization techniques for enhancing the performance and reliability of these applications.

Furthermore, future research relies on the objective of data management and storage. The operation of making cost-effective adoption applications is also considered in further researches. Besides, by using statistical-based methods, IoT can help to predict the upcoming status of COVID-19. With this technology's suitable achievement, researchers, doctors, government, and academicians can improve fighting the COVID-19.

References

1. Statistics of Corona Virus meters for increasing infected people. https://www.worldometers.info/coronavirus/
2. Corona Virus (COVID-19) Deaths. https://ourworldindata.org/covid-deaths
3. Agi, K.: Ph.D. President & CEO of Sensor Comm Technologies, co-authored this White Paper on Remote Patient Monitoring for COVID-19
4. Shahrestani, S.: Internet-of-things and smart environments, Assistive Technologies for Disability, Dementia, and Aging. Springer (2017)
5. Schwartz, E.: The Global Health Worker Shortage: 7 Numbers to Note. https://www.projecthope.org/the-global-health-worker-shortage-7-numbers-to-note/02/2020/ (2020)
6. Yeole, A., Kalbande, D.R.: Use of Internet of Things (IoT) in Healthcare: A Survey, Conference: the ACM Symposium (2016)
7. Maxim, C., Zubair A.B.: Internet of Things (IoT): research, simulators, and testbeds. IEEE Inter. Things J. **99**(1–1)
8. Gabriele, D., Stefano, F.: Simulation of The Internet of things, Conference: 2016 International Conference on High Performance Computing & Simulation (HPCS) (2016)
9. Galletta, A., Carnevale, L., Bramanti, A.: Maria Fazio. An innovative methodology for big data visualization for telemedicine, IEEE Trans. Ind. Infor. (2018)
10. Radovic, M., Ghalwash, M., Filipovic, N., Obradovic, Z.: Minimum redundancy maximum relevance feature selection approach for temporal gene expression data. BMC Bioinf. **18**(9) (2017)
11. Kuang, S., Davison, B.D.: Learning word embeddings with chi-square weights for healthcare tweet classification. Appl. Sci. **7**, 846 (2017)
12. Khoie, M.R., Sattari Tabrizi, T., khorasani, E.S., Rahimi, S., Marhamati, N.: A hospital recommendation system based on patient satisfaction survey. Appl. Sci. **7**, 966, 2017
13. Dziak, D., Jachimczyk, B., Kulesza, W.J.: IoT-based information system for healthcare application: design methodology approach. Appl. Sci. **7**, 596 (2017)

14. Sundaravdivel, P., et al.: Everything you wanted to know about smart health care: evaluating the different technologies and components of the internet of things for better health. IEEE Consum. Electron. Mag. **7**(1), 18–28 (2018)
15. Favaloro, E.: Recommendations for minimal laboratory testing panels in patients with COVID-19: potential for prognostic monitoring, seminars in thrombosis and hemostasis (2020)
16. Olutosin, T., Absalom, E.E.: Smart healthcare support for remote patient monitoring during covid-19 quarantine. Infor. Med. Unlocked **20**, 100428 (2020)
17. Rameswari, R., Divya, N.: Smart health care monitoring system using android application: a review. Inter. J. Recent Technol. Eng. (IJRTE) **7**(48) (2018)
18. Senthamilarasi, C., Jansi Rani, J., Vidhya, B., Aritha, H.: A smart patient health monitoring system using IOT. Inter. J. Pure Appl. Math. **119**(16), 59–70 (2018)
19. Mora ID, H., Gil ID, D., Terol D, R.M., Azorín D, J., Szymanski, J.: An Iot-based computational framework for healthcare monitoring in mobile environments, sensors (2017)
20. Imteaj, A., Muhammad, K.H.: A smartphone based application to improve the health care system of bangladesh
21. Yi, W.-J., Saniie, J.: Patient centered real-time mobile health monitoring system. E-Health Telecomm. Syst. Netw. **5**, 75–94 (2016)
22. Jeong, J-S., Han, O., Yen, Y.Y.: A design characteristics of smart healthcare system as the IoT application. Indian J. Sci. Technol. **9**(37) (2016)
23. Allam, Z., Jones, D.S.: On the coronavirus (COVID-19) outbreak and the smart city network: universal data sharing standards coupled with artificial intelligence (AI) to benefit urban health monitoring and management. Healthcare (Basel) **8**(1) (2020)
24. Mshali, H., Lemlouma, T., Moloney, M., Magoni, D.: A survey on health monitoring systems for health smart homes. Inter. J. Ind. Ergon. Elsevier **66**, 26–56 (2018). https://doi.org/10.1016/j.ergon.2018.02.002
25. Singh, R.P., Javaid, M., Haleem, A., Suman, R.: Internet of things (IoT) applications to fight against COVID-19 pandemic. Diabetes Metab Syndr. **14**(4) 521–524 (2020)
26. Lu, J., Zhang, S.: E-health web application framework and platform based on the cloud technology, MSC thesis, School of Health and Society Department Design and Computer Science (2013)
27. Mwaffag, O., Nesreen, O., Mohammad, A.A., Yousef, E., Rudaina, B.: An IoT-based framework for early identification and monitoring of COVID-19 cases. Biomed Signal Process Control. (2020)
28. Alshraideh, H., Otoom, M., Al-Araida, A., Bawaneh, H., Bravo, J.: A web based cardiovascular disease detection system. J. Med. Syst. **39**(10), 122 (2015)
29. Talakala, S., Hari Krishna, M.: Instantaneous health care monitoring system d smart phone. J. Electro. Control Syst. Control Instrum. Eng. **2**(2) (2016)
30. Raj, G., Prakash, N.R., Randhawa, J.S.: IoT based EMG monitoring system. Inter. Res. J. Eng. Technol. (IRJET) **4** (2017)
31. Bhayani, M., Patel, M., Bhatt, C.: Book: Internet of Things (IoT): in a way of smart world. Proce. Inter. Congr. Infor. Commun. Technol. Adv. Intell. Syst. Comput. Book Series **438**, 343–350 (2016)
32. Prajoona, V., Tariq Ahmed, B.B., Baabood, A.H.O.: IOT based health monitoring system. J. Crtiticl Rev. **7**(4), 739–743 (2020)
33. Livinsa, M.Z., Valantina, G.M., Premi, M.S.G.: Health monitoring systems in smart environments. J. Pharm. Sci. Res. **11**(9), 3130–3132 (2019)
34. Teng, X-F., Zhang, Y-T., Poon, C.C.Y., Bonato, P.: Wearable medical systems for p-health. IEEE Rev. Biomed. Eng. 62–74 (2008)
35. Omoogun, M., Ramsurrun, V., Guness, S., Seeam, P., Bellekens, X., Seeam, A.: Critical patient eHealth monitoring system using wearable sensors. In: 2017 1st International Conference on Next Generation Computing Applications (NextComp), IEEE, pp. 169–174 (2017)
36. Liang, T., Yuan, Y.J.: Wearable medical monitoring systems based on wireless networks: a review. IEEE Sens. J. **16**(23), 8186–8199 (2016)

37. Haahr, R.G., Duun, S.B., Toft, M.H., Belhage, B., Larsen, J., Birkelund, K., Thomsen, E.V.: An electronic patch for wearable health monitoring by reflectance pulse oximetry. IEEE Trans. Biomed. Circuits Syst. **6**(1), 45–53 (2011)

38. Tang, Y-W., Schmitz, J.E., Persing, D.H., Stratton, C.W.: The laboratory diagnosis of COVID-19 infection: current issues and challenges. J. Clinical Microbiol. (JCM) (2020)

39. Islam, M.M., Rahaman, A., Islam, M.R.: Development of smart healthcare monitoring system in IoT environment. SN Comput. Sci. **185** (2020)

40. COVID-19 Technical specifications for procurement of oxygen therapyand monitoring devices, World Health Orgnization, MedDev/TS/O2T.V1 (2020)

41. El-Din, D.M., Hassanein, A.E., Hassanien, E.: E-Quarantine: a smart health system for monitoring coronavirus patients for remotely quarantine. arXiv:2005.04187

42. Noun, A., Hemanvtha, K.: An approach to monitor coronavirus using IoT and machine learning (2020)

43. Dan, C., Irina, I., Gabriela, N.: IoT system in diagnosis of Covid-19 patients. Infor. Econ. **24**, 75–89 (2020)

44. Kumar, K., Kumar, N., Shah, R.: Role of IoT to avoid spreading of COVID-19. Inter. J. Intell. Netw. **1**, 32–35 (2020)

45. Swati, S., Chandana, M.: Application of cognitive Internet of Medical Things for COVID-19 pandemic. Diabetes Metab. Syndr. **14**(5), 911–915 (2020)

46. Yi, Y., Lagniton, P.N.P., Ye, S., Li, E., Xu, R-H.: COVID-19: what has been learned and to be learned about the novel coronavirus disease. Inter. J. Biol. Sci. **16**(10), 1753–1766 (2020)

47. Darwish, A., Hassanien, A.E., Elhoseny, M., Sangaiah, A.K., Muhammad, K.: The impact of the hybrid platform of Internet of things and cloud computing on healthcare systems: opportunities, challenges, and open problems. J. Ambient Intell. Hum. Comput. **10**(10), 4151–4166 (2019)

COVID-19 and Water Resources Nexus: Potential Routes for Virus Spread and Management Using Artificial Intelligence Techniques

Hend S. Abu Salemⓘ, **Mahmoud Y. Shams**ⓘ, **Aboul Ella Hassanien**ⓘ, **and Ahmed M. Nosair**ⓘ

Abstract The new Coronavirus (COVID-19) pandemic has cast a shadow over the entire world and affected the universe in all economic, environmental, industrial, and agricultural aspects. Accordingly, the study of the potential routes of virus spreading is crucial to limit the infection. This chapter presents a comprehensive survey of most past and current studies that dealt with the causes of the spread of the new COVID-19, mostly surface water and groundwater. The threat is commonly associated with the direct use of contaminated surface or groundwater by sewage sources arose mainly from the mixing of clean water with sewage through several pathways. The second wave of COVID-19 is currently invading the world, especially with the emerging new virus strains that invaded Britain in December 2020, having double the infectivity of the COVID-19 of the first wave. Accordingly, a water management system is required to monitor the increased transmission rate by developing biosensors and IoT to detect and estimate the SARS-CoV-2 in wastewater and drinking water. Besides, the integration of IoT and artificial intelligence (AI) methodologies could examine how effectively the AI could monitor the outbreaks to help decision-makers curb the infection. This work will also shed light on increasing public awareness of the virus's risk spread through distinct contaminated surface and groundwater paths, especially when consumed directly without disinfection procedures. Additionally,

H. S. Abu Salem (✉)
Geology Department, Faculty of Science, Cairo University, Cairo, Egypt
e-mail: hendsaeed@cu.edu.eg

M. Y. Shams
Faculty of Artificial Intelligence, Kafrelsheikh University, Kafr El Shaikh 33511, Egypt
e-mail: mahmoud.yasin@ai.kfs.edu.eg

A. E. Hassanien
Faculty of Computers and Artificial Intelligence, Cairo University, Cairo, Egypt
e-mail: aboitcairo@cu.edu.eg

A. M. Nosair
Environmental Geophysics Lab (ZEGL), Geology Department, Faculty of Science, Zagazig University, Zagazig, Egypt

H. S. Abu Salem · M. Y. Shams · A. E. Hassanien · A. M. Nosair
Scientific Research Group in Egypt (SRGE), Cairo, Egypt

© The Author(s), under exclusive license to Springer Nature Switzerland AG 2021
A. E. Hassanien et al. (eds.), *The Global Environmental Effects During and Beyond COVID-19*, Studies in Systems, Decision and Control 369,
https://doi.org/10.1007/978-3-030-72933-2_2

19

a case study in the East Nile Delta is discussed to show how the contamination of surface water represents a potential route for virus transmission through the direct use of contaminated groundwater, where a potential wastewater flow paths were recorded from a heavily polluted drain to the nearby wells. We recommend a future work plan for continuous monitoring of the surface water, groundwater, and wastewater, to detect the presence or absence of the SARS-CoV-2 virus.

Keywords COVID-19 · Surface water · Groundwater · Second wave · Artificial intelligence · Internet of things

1 Introduction

Water resources have widely been contaminated by several pathogens of a significant threat to human health. The most common pathogens are viruses and pathogenic protozoa. The threat is commonly associated with direct use of contaminated surface or groundwater, where almost half of all waterborne disease outbreaks are linked to contaminated water [17]. The contamination sources arose mainly from the mixing of clean water with sewage through several pathways.

The current global pandemic caused by the novel coronavirus 2 (SARS-CoV-2, known as COVID-19) is proven to be transmitted through inhalation of respiratory droplets and person to person contact routes [48, 90]. Other forms of transmission include airborne transmissions such as aerosols and atmospheric particulates, which may provide significant pathways for COVID-19 transmission [32]. Moreover, the sewage and human excreta could be regarded as a potential route for transmitting COVID-19 into the environment where the SARS-CoV-2 virus is found in the feces and urine of patients that experience severe acute respiratory syndrome [45, 52, 53, 67, 80, 94, 96, 99]. Patients of asymptomatic status [82] and treated patients with no further sign of the symptoms [96] also have SARS-CoV-2 virus in their feces. The virus could remain viable for days and show infectivity, which poses concerns to the scientific community [4, 8, 16, 32, 64, 72, 88]. The SARS-CoV-2 virus is also detected in both influents and treated effluents of wastewater treatment plants [34, 72, 74].

This article focuses on the different pathways that COVID-19 can contaminate surface and groundwater and the management and control of water resources used during the virus outbreak. Besides, artificial intelligence (AI) methodologies will be discussed to show how effectively it can monitor the outbreak. Besides, this work will shed light on increasing the public awareness of the risk of the virus spread through sewage-contaminated surface and groundwater, especially when consumed directly without disinfection procedures.

2   Background About Viruses in Water with Particular Emphasis on COVID-19

Pathogenic organisms such as viruses and protozoa pose a severe threat to human health, especially when they are diffused in the surface or groundwater. Most water-borne disease outbreaks are linked to contaminated groundwater [17], where several sources were found to be responsible for the spread of pathogens in the surface and groundwater. Surface water could be contaminated by the direct discharge of untreated or partially treated wastewater in rivers, canals, lakes, and the manure run-off from agricultural lands. In contrast, groundwater is mainly contaminated due to the migration of sewage from septic tanks or sewer leakage.

Unlike other pathogens, the small size of viruses prevents them from being filtrated from the soil as pathogens of larger sizes [46]. Besides, the fate of pathogenic protozoa in the soil is ambiguous, while the information about the removal of viruses during travel in the soil is well-known [33]. The transport of viruses in porous media is controlled by (i) the mechanisms and modeling of virus sorption, (ii) the virus survival and factors affecting virus inactivation in the natural environment, and (iii) the mechanisms of virus transport in porous media [46]. Accordingly, information about the solution chemistry, virus properties, soil properties, temperature, association with solid particles, and water content is useful for modeling the virus sorption, survival, and transport in porous media [46].

3   Potential Routes of COVID-19 Spread to Humans via Water

Various potential routes provide viral transmission to humans, such as contaminated surface and groundwater in addition to crops grown in contaminated soil. The following context will give information on these routes.

3.1   Sewage-Associated Transmission of COVID-19

The causes that could be responsible for the spread of COVID-19 in surface and groundwater water arise from human feces' contamination [91]. Disposing municipal and medical effluents into surface water bodies (drains, or canals) represents the main route for contaminating surface water by pathogens like COVID-19. On the other hand, the existence of different groundwater pathogens is less common than in surface water. However, very few occurrences make the water unsuitable for drinking purposes [6]. In groundwater systems, the water flow in the aquifer material helps remove and reduce the virus activity. The filtration and disinfection procedures in most public drinking water systems eliminate and/or inactivates viruses. Despite the low risks, the question has arisen about the vulnerability to COVID-19 through direct use of untreated public groundwater.

The detection of SARS-CoV-2 RNA in hospital sewage and public wastewater [4, 55, 59, 88] pose a substantial threat to the environment. The sewage water could diffuse to the surface and groundwater through different paths. Even though the persistence of SARS-CoV-2 in wastewater remains uncertain [35], the occurrence of the SARS-CoV-2 RNA in untreated and treated wastewater, as well as the presence of the infectious SARS-CoV-2 in stool samples [94, 96, 105], bears potential risk for virus spreading in the environment through those sources [10, 31, 72].

Even though SARS-CoV-2 loses its infectivity very quickly in wastewater, according to the work of Venugopal et al. [88], and Annalaura et al. [10], waning the virus transmission through virus-contaminated water, the spreading risk arises from the historical records of epidemic outbreaks from the consumption of contaminated water especially in low-income countries [70]. These outbreaks were mainly related to diarrhea and gastroenteritis.

Other sources that contribute sewage contamination to the environment include the leakage from sewers to shallow aquifers in urban areas. This leakage could arise from old drainage infrastructure (Fig. 1). Additionally, sewers could damage and crack due to earthquakes or intense penetration of tree roots (Fig. 1). The untreated wastewater is often discharged into the environment [3], which accidentally could find its way to the groundwater through porous soils [65].

On the other hand, in rural and peri-urban areas, the poor lining of septic tanks and the possible proximity of pit latrine sanitation systems to groundwater source increases the risk of sewage leakage to the surrounding environment and bear risks to

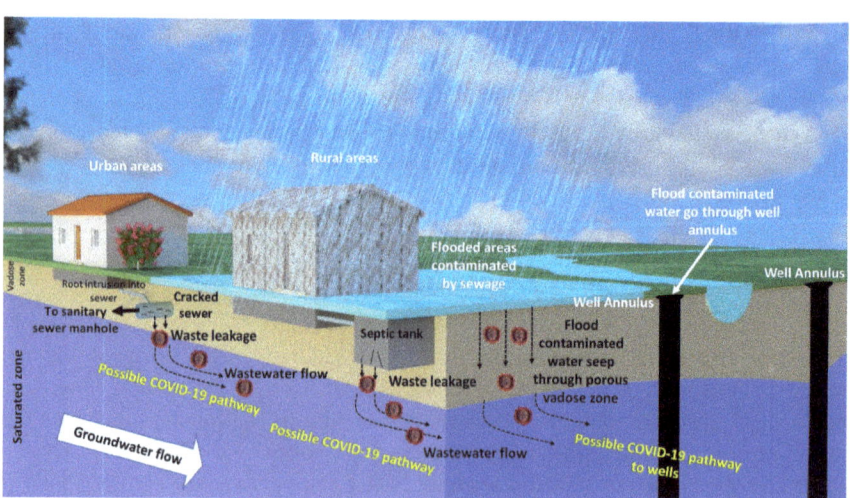

Fig. 1 Sewage-associated transmission of COVID-19 through aged broken sewers, during flooding, and around unproperly sealed well annuli

nearby water resources (Fig. 1, [2]). Additionally, the unplanned and open defecation close to surface water also promotes virus spreading [12, 14, 89].

3.2 Effect of Flooding on Virus Spreading

Flooding, a natural hazard driven by heavy rains, is reported in urban and rural areas. Sewage overflows usually accompany excessive flooding in urban areas. Sewage-associated transmission of COVID-19 in flooded areas were studied by Han and He [32], where sewage overflows appear in communities serviced by combined sewer systems. The sewage overflows could be exacerbated by increasingly violent conditions of rising rainfall caused by climate change, urban expansion, and aging of sewers and storm infrastructure in many places that contribute waste into living areas and in the vicinity of water bodies and beaches (Fig. 2a, b) [7, 95]. In addition, during severe flooding events, the combined sewer overflows could be worsened by the spontaneous loads of wipes, masks, and gloves flushed down the drain or improperly disposed of at the roadside during the pandemic, obstructing the flow in drainage pipes and sewage pumping stations [60].

The studies that postulated the sewage-associated transmission of COVID-19 [62, 71, 100] confirmed their hypothesized transmission routes by tracking the overflows from broken pipes during heavy rainfall events. Moreover, the risks of virus transmission via sewage-contaminated aerosols were documented in earlier outbreaks of SARS 2003. These aerosols were generated in large quantities in drainage pipes "the vertical soil stacks" when flushing toilets. Yu et al. [101] recorded huge numbers of virus-laden aerosols produced by the hydraulic action in vertical soil stacks when toilets were flushed. The virus-laden aerosols were produced when extremely high concentrations of the SARS-associated coronavirus found in patients' feces and urine, accompanied by aerosolization due to hydraulic action inside the vertical soil stacks (Fig. 3, [101]). Additional risks arise from the bathroom exhaust fans

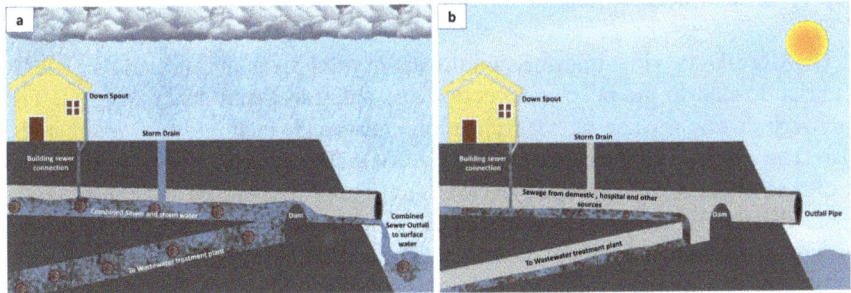

Fig. 2 Combined sewer systems in **a**) overflows during wet weather as a potential route of virus shedding to surface water, **b**) absence of sewer overflows in dry weather (no chance to contaminate surface water)

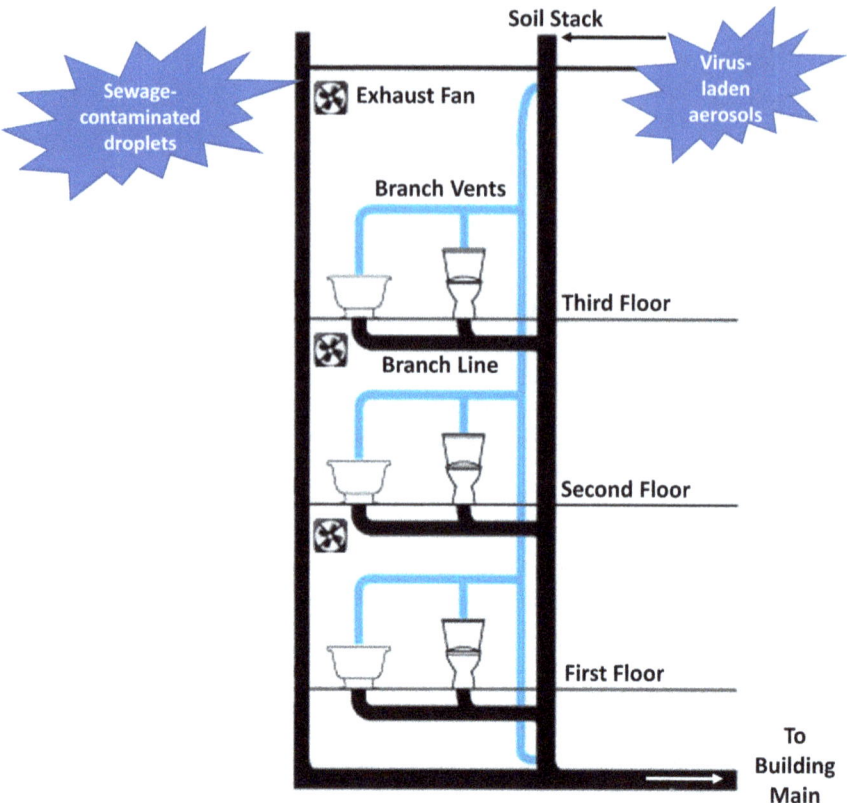

Fig. 3 Virus laden aerosols inside the vertical soil stacks after flushing toilets and sewage-contaminated droplets when bathroom exhaust fans working in closed doors conditions

working in closed doors conditions, where the sewage-contaminated droplets were drawn from drainage pipes via dry-out floor drain traps. Further, transporting those droplets to apartments several floors away through the building's light well (Fig. 3, [30]).

The COVID-19 virus transmission threats in rural areas during flooding events could be related to the poor construction and aging of septic tanks (Fig. 1). The discharge of raw sewage drawn from septic tanks into nearby drains and canals. Also, The sewage-associated transmission of COVID-19 during flooding could also be of potential risk if the groundwater wells have poor well design and construction, where the poorly constructed well annuli could have enough porosity to let the sewage-contaminated flooding to enter into the aquifer (Fig. 1).

Accordingly, a proper sanitary seal around the well casing is essential to block contaminants that might migrate from the land surface down the well annulus to the water table, bypassing the unsaturated zone that contributes to the natural purification of groundwater.

In conclusion, the dependence on combined sewer systems in urban areas increases the risk of the COVID-19 virus spreading during flooding. The shortcuts of exposure to overflowed sewage and non-dispersed human excreta during urban flooding events could be obtained by improving wastewater infrastructure, considering the combined sewer separations to minimize pathogen risks transmission via sewage overflows during epidemics. In rural and peri-urban areas in low-income countries, the potential risk of community spread of COVID-19 increases as the dependence on water resources' direct use increases.

3.3 COVID-19 Spread Related to Water Pollution by Persistent Organic Pollutants (POPs)

The persistent organic pollutants (POPs) are widely spread in the environment through sediments, animals, vegetation, rivers, and groundwater due to the widespread anthropogenic activities [36, 76, 97]. They include polychlorinated biphenyls (PCBs), polycyclic aromatic hydrocarbons (PAHs), organo-chlorines, organophosphates, carbamates, dioxins, heavy metals, methylmercury, and endocrine disrupters. They could find their route to animals and humans through diet [84]. They produce many biological effects on wildlife and humans, including immunological, mutagenic, and reproductive alterations [102].

The relation between chronic and infectious diseases related to contaminant exposure has been studied in recent decades [19, 20, 54]. The main effects of these pollutants are the capability of depressing the immune system of humans and animals and thus might increase the impacts of pathogens, such as the global coronary disease (COVID-19) [24]. Accordingly, the long-term exposure to anthropogenic pollutants like POPs and PCBs through diet and drinking water could have an important role in COVID-19 spreading [87].

3.4 Effect of Sewage Sludge Applications in Agriculture on COVID-19 Spreading

The sewage sludge is considered one of the most common fertilizers extracted during the wastewater treatment process. The viruses detected in septic tanks, sewage sludges, and wastewater could have ecological and health impacts of these sources that were disseminated into the environment. Accordingly, a question about the potentiality of the virus spreading as a result of the application of sewage sludges produced during the COVID-19 epidemic should be answered. Viruses dispersed from the wastewater and sludge applications to fields could be deposited on crops and soil where they are likely to survive [50]. As a result, the consumption of contaminated garden products and the possible contamination of surface and groundwater due to virus-laden amendments denote public health risk. The fate of these viruses, particularly in plants, must therefore be known especially after the steady increase of reuse of treated and untreated wastewater in agriculture. It is worth mentioning that the presence of a viral genome does not indicate infectious viral particles [50].

The survival of viruses on plants irrigated with sewage ranges from a few days to nearly four weeks, depending on conditions [13, 85]. Another factor that influences the survival time of viruses on plants after irrigation is the nature of cultivated vegetables, the initial level of contamination, and the temperature associated with sunlight [50]. Accordingly, attention should be driven to the use of untreated wastewater and sewage sludge in agriculture, where strict disinfection treatment should be considered in accordance with the regulations. In addition, a storage period could be specified for the sewage sludge beyond which the virus is inactivated.

3.5 Fecal Coliform Bacteria and Artificial Sweeteners in Surface and Groundwater

The presence of fecal coliform bacteria and artificial sweeteners, such as acesulfame-K, in surface and groundwater, represents a good indicator of sewage contamination. The direct use of these waters, without disinfection, poses a significant risk to COVID-19 spreading, especially in low-income countries, for drinking and domestic purposes. Several workers have studied the use of artificial sweeteners (known as exogenous human biomarkers) to trace sewage pollution in surface and groundwater in the last few years [51, 73, 79, 86]. These biomarkers as integrated wastewater-based epidemiology (WBE) programs could be an invaluable tool in predicting the COVID-19 pandemic [69]. The quantification of biomarkers could indicate to measured viral concentrations where higher loads relative to population indicate a viral outbreak [66, 78].

3.6 Microplastics as a Possible Factor in COVID-19 Spreading

Microplastics have been widely known in recent years as a common pollutant to surface water [37, 93, 103, 104].

A load of microplastic pollution is increasing in the current pandemic due to extensive use of personal protective equipment (PPE) including face masks, disposable gloves, and disinfectant wipes, which are often made of single use plastic [9, 26].

As a result of data limitation on the environmental sustainability of SARS-CoV-2, the basic understanding of its persistence can be explained by the results of studies on other coronaviruses such as SARS CoV-1 and MERS CoV. The survival of the SAR-CoV-2 virus is phylogenetically similar to that of SAR-CoV [26]. The survival rate with a viral titer of 10^5 on plastic is from 4–5 days at room temperature [47]. The microplastic pollution of water resources is of significant concern because the plastic particles allow colonization of several pathogenic microorganisms (such as bacteria, viruses, fungal filaments, and spores) and allow disease transmission through direct use of contaminated water for drinking purposes.

4 Case Study- Expected Flow Paths for COVID-19 Transport Through Porous Media

Wastewater may function as a source for virus spreading if this water is hydraulically linked to fresh groundwater sources and if significantly polluted with the hospital or sewage effluents. The infection potential may also arise during the wastewater treatment processes or from direct contact with polluted water. As previously discussed, the groundwater, especially that exposed to seepage from wastewater, is responsible for endemic enteric diseases [15].

Groundwater in many urban areas can easily be contaminated by enteric viruses of extremely small sizes, which easily transmitted to considerable distances through the infiltrated water [21, 28, 29]. Viruses found in different wastewater types can be easily transported into groundwater and surface water if they are directly thrown in surface water streams or infiltrated into the groundwater through porous soils [46]. In many countries, the different contaminants could be transported from the unmanaged sewage systems into groundwater, especially at shallow depths, which cause many hydrological and environmental problems [98]. The contamination problem could be exacerbated when using wastewater for irrigation due to the lack of water resources.

The Nile Delta of Egypt is taken as an example to illustrate the probability of groundwater contamination by enteric viruses like SARS-CoV-2. This area is dominated by agricultural lands where the surface is dissected by several unlined drainage networks transporting different types of wastes (agriculture return flow, industrial, hospitals, and others). This wastewater could find their path to shallow aquifers via porous soil posing risks to the water resource [18]. Porous media such as sands, gravels, silty sands, and sandy clays are considered potential flow paths for contaminants from polluted surface water to shallow groundwater aquifers. The contaminant's flow is enhanced under high pumping rates and head potential between surface water and groundwater levels. The eastern part of the Nile Delta is an excellent example for this case, where it is dissected by many unlined open drains such as El Qalyubiya, Bilbies, and Bahr El Baqr (Fig. 4a). These drains transport a huge load of several contaminants for long paths. These drains and the neighboring shallow groundwater aquifers are previously studied by many authors where they recorded high contamination potential in the drains and some nearby wells [1, 75, 81, 106]. The main contaminants are heavy metals and pathogens like fecal and total coliform bacteria, reflecting seepage from these drains to the shallow groundwater aquifers through the permeable soils. Abu Salem et al. [1] detected fecal and total coliforms in

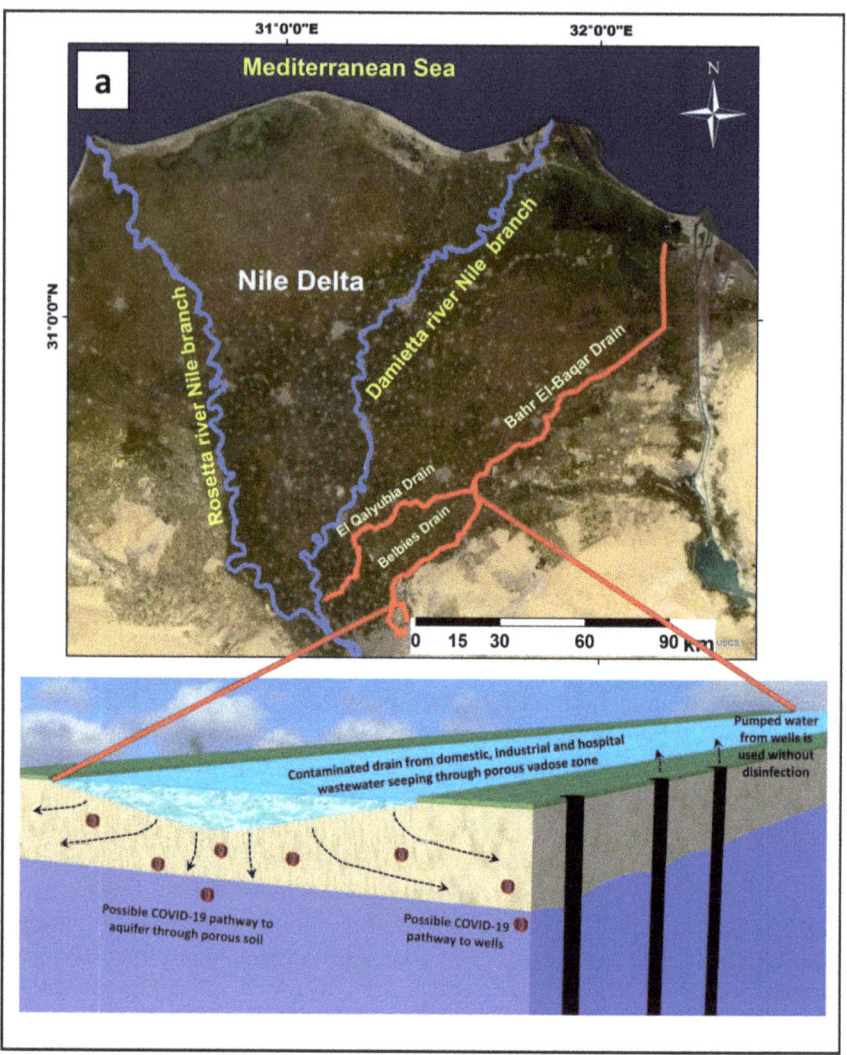

Fig. 4 a) Potential flow paths of wastewater from highly polluted drain to shallow aquifer and nearby wells. **b**) Increased infection cases in Belbies center in the first wave of COVID-19 (Daily reports of the Egyptian Ministry of Health and Population, 2020)

the Belbies drain and some shallow groundwater wells near the drain. They concluded that flow pathways are transporting contaminants from the Belbies drain to those wells through the porous soil that line the drain.

The wells are represented mainly by hand pump wells drilled at shallow depths ranging from 15 to 30 m and representing the main source of drinking and irrigation purposes. The contamination of the shallow groundwater by fecal coliform bacteria could give insights about the possibility of transporting SARS-CoV-2 to

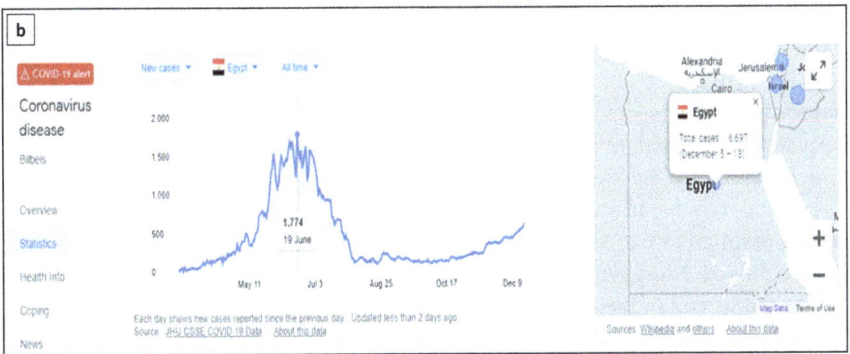

Fig. 4 (continued)

the groundwater endangering the consumers. This situation could be further worsened when these shallow groundwater wells are over-pumped, promoting the flow to the aquifer. The infections by COVID-19 were recorded in many villages and cities around and near to the Belbies drain, especially the Belbies city as well as several nearby villages in the first wave of the epidemic, confirming our expectation (Fig. 4b, Daily reports of the Egyptian Ministry of Health and Population, 2020). We recommend a future work plan for continuous monitoring of the groundwater wells, especially the shallow ones, to detect the presence or absence of the SARS-CoV-2 virus. The constant tracking of wastewater is vital for mitigating infectious diseases in water [11].

5   Water Management and Control During COVID-19

Since sensitive equipment and facilities in water systems are necessary for testing the presence of SARS-CoV-2 and other enteric viruses, the need for more cheap devices is crucial. The most sensitive technique for detecting and quantifying the virus mRNA is the RT-PCR (reverse transcription-polymerase chain reaction) test, which helps identify gene fragments of the SARS-CoV-2 virus. The method of detecting viruses in wastewater and drinking water embraces fast, inexpensive, sensitive, resistance to false-positive results, and full automation has not yet been established [57].

Some studies have recently reported using inexpensive systems to detect SARS-CoV-2 in wastewater, such as paper-based and biosensor devices [56, 58]. The development of techniques to track the spread of coronaviruses in water is essential, especially if there are second or more waves. Kitajima et al. [48] are currently working on to develop test methods for monitoring SARS-CoV-2 genetic material in sewage. The study is based on an estimation of the disease frequency in communities, identifying areas where tests were done, and predicting the possibility of second wave of infection. Yet, this approach is not an alternative to human testing. In addition to economic agents, the implementation of methods for the detection of viruses in the aquatic environment can come up against several obstacles, including the considerable dilution of the sample, the influence of the environmental matrix on the analytical results, and the mutagenic variability of viruses [57].

An alternative method is to use artificial intelligence (AI) methodologies in monitoring the spread of the virus using IoT. The biosensors collect the data from different sources that may be affected by the COVID-19 virus. These data represent the raw material enrolled in the AI approaches to detect, recognize, classify, and regress the results obtained. Moreover, the AI comes to provide a suitable decision to track and monitor the spread of the virus in different regions.

Recently, water management control issues have become significant directions to all research, especially in the current COVID-19 pandemic. The appropriate water management control system's appropriate foundations are the crucial achievement for designing optimal water management to reduce water consumption for increasing the water infrastructure utilization. In this work, we believe that it is of extreme significance to merge artificial intelligence (AI) with the Internet of Things (IoT), to attain the superlative possible water management control performance during this pandemic.

Artificial Intelligence depending on optimization approaches, as in Firefly algorithm (FA) is considered as one of the most powerful improvement strategies that is dependent on swarm intelligence [5, 22, 23, 27, 38, 39]. Deep Learning (DL) approaches presented by Shen [77] offered a particular framework and transdisciplinary development update about water relevance. DL approach is utilized to handle information problems, including enhancement of productivity and increased knowledge. This rising sub-discipline has exposed the DL techniques counting the correlation-based analysis and the reverse of the extracted network. Moreover, DL approaches able to utilize the enrolled features for condition neurons to handle as many as possible issues related to water sciences as presented in [37–44, 40].

6 IoT in Water Management Control

IoT plays an essential role in managing and control the water during the recent COVID-19 pandemic. Such smart water management control systems have just instigated to gain force as real-time water utilities started to use the enrolled water data that are acquired and stored to be statistically analyzed using machine learning (ML) and Deep Learning (DL) approaches to effectively maintain and handle water scarcity [49].

In IoT water control management system, data can interconnect with each other devices including water utilities and smart meters, in the presence of AI, ML, and DL to analytically design the suitable architecture to produce the right decision regarding specific actions to track and monitor the water supply in the system [83].

Furthermore, AI can also forecast characteristic patterns in the water management control system during the COVID-19 pandemic. Because of the complexity involved in the water management control system contains pumping stations, reservoirs, and consumer services, the forecasting process are more accurate enough to help to monitor the utilities of water that might be raised in the times of water leakage and peak consumption [68]. In Nasser et al. [63], the authors presented a two-layer water Prediction System based on Long Short-Term Memory (LSTM) Neural Networks to boost the water management control system.

As investigated in Fig. 5, the general schematic diagram for water management control system using IoT biosensors and AI approaches to track and monitor the existence and spread of the COVID-19 virus is presented as follows:

i.   Water data collection or groundwater data or the infected location.
ii.   Biosensors are applied to measure the virus concentration.
iii.   The data are collected and transmitted using IoT or Mobile application.
iv.   The preprocessing stage include the following:

    a.   Segmentation or clustering the input data.
    b.   Extract the features and/or select the most significant features.
    c.   Decide the quality control system based on a specified threshold or applied filter.
    d.   The extorted features are classified, regressed using a supervised machine learning algorithm.

v.   The matching scores are produced to provide a suitable decision.

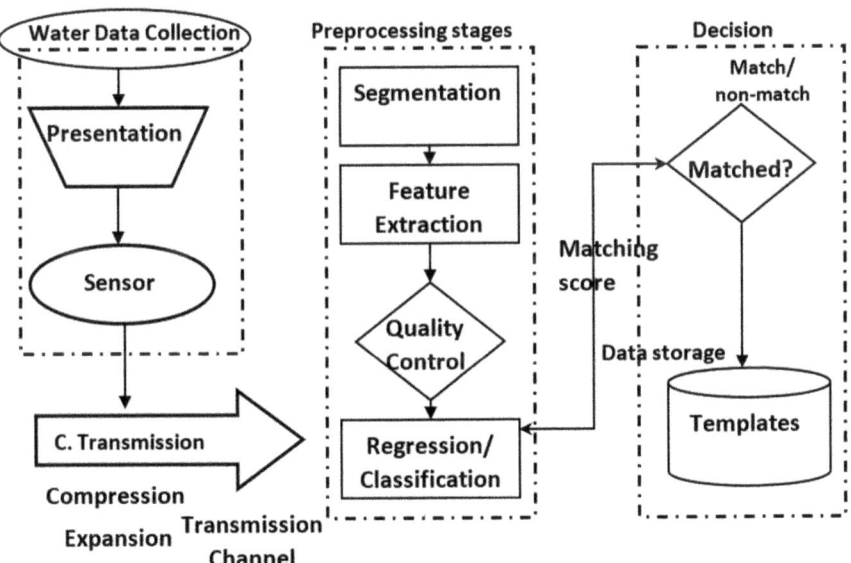

Fig. 5 General schematic diagram of the AI-based water management control system during COVID-19 pandemic

vi.   Depending on the matching algorithm and the applied threshold or filter, a decision of matched or classified data is stored in the database.
vii.   The data are trained, tested, and validated using ML approaches.

7   Conclusions

This article focuses on the different pathways that the novel coronavirus 2 (SARS-CoV-2, known as COVID-19) can contaminate surface and groundwater. The threat is commonly associated with the direct use of contaminated surface or groundwater by sewage sources. The contamination sources arose mainly from the mixing of clean water with sewage through several pathways.

The current global pandemic caused by the novel coronavirus 2 (SARS-CoV-2) is proven to be transmitted through inhalation of respiratory droplets and person to person contact routes. Other forms of transmission include airborne transmissions such as aerosols and atmospheric particulates, which may provide significant pathways for COVID-19 transmission. Moreover, the sewage and human excreta could be regarded as a potential route for transmitting COVID-19 into the environment where the SARS-CoV-2 virus is found in feces and urine of patients that experience severe acute respiratory syndrome as well as in patients of asymptomatic status and treated patients with no further sign of the symptoms. The virus viability in stool and urine samples lasts for days showing infectivity and posing concerns to the scientific community. Accordingly, other methods that contaminate water resources by sewage is the aging of sewer infrastructure, the unproper lining of septic tanks, and the proximity of pit latrine sanitation systems to groundwater sources.

Additionally, the contamination of surface and groundwater could be induced by intense flooding events especially when the combined sewer systems dominate. Other indirect transmission routes through contaminated water by persistent organic pollutants (POP) and the presence of artificial sweeteners in water resources are also discussed. The POP compounds have a detrimental effect on the human immune system, leading to pathogenic infection's high susceptibility. The presence of artificial sweeteners in water resources indicates the contamination by sewage, representing a proven route for COVID-19 transmission. Furthermore, the microplastic pollution of water resources could be considered as a potential route for virus transmission due to the capability of plastic particles to allow colonization of several pathogenic microorganisms (such as bacteria, viruses, fungal filaments, and spores) and allow disease transmission through direct use of contaminated water for drinking purposes.

A case study in the East Nile Delta is discussed to show how the contamination of surface water represent a potential route for virus transmission through the direct use of contaminated groundwater, where possible wastewater flow paths were recorded from a heavily polluted drain to the nearby wells. Furthermore, the second wave of COVID-19 is currently invading the world, especially with the emerging new virus strains that invaded Britain in December 2020, having double the infectivity of the COVID-19 of the first wave. Accordingly, a water management system is required to monitor the increased transmission rate by developing biosensors and IoT to be used in the detection and estimation of the SARS-CoV-2 in wastewater and drinking water. The integration of IoT and artificial intelligence (AI) methodologies could examine how effectively the AI could monitor the outbreaks to help decision-makers curb the infection.

We recommend a future work plan for continuous monitoring of the surface water, groundwater, and wastewater, to detect the presence or absence of the SARS-CoV-2 virus.

This work will also shed light to increase the public awareness of the risk of the virus spread through different paths in contaminated surface and groundwater, especially when consumed directly without disinfection procedures.

Acknowledgements The authors would like to thank Mr. Ahmed O. Hamad for drawing the figures for this article.

References

1. Abu Salem H., Gemail, K.H.S., Nosair, A.M.: A multidisciplinary approach for delineating wastewater flow paths in shallow groundwater aquifers: a case study in the southeastern part of the Nile Delta, Egypt. J. Contaminant Hydrol. (2020) https://doi.org/10.1016/j.jconhyd.2020.103701
2. Adelodun, B., Ajibade, F.O., Ibrahim, R.G., Bakare, H.O., Choi, K.S.: Snowballing transmission of COVID-19 (SARS-CoV-2) through wastewater: any sustainable preventive measures to curtail the scourge in low-income countries? Sci. Total Environ. **742**, (2020)
3. Adelodun, B., Odedishemi, F., Segun, M., Choi, K.-S.: Dosage and settling time course optimization of Moringa oleifera in municipal wastewater treatment using response surface methodology. Desalin. Water Treat. **167**, 45–56 (2019). https://doi.org/10.5004/dwt.2019.24616
4. Ahmed, W., Angel, N., Edson, J., Bibby, K., Bivins, A., O'Brien, J.W., Choi, P.M., Kitajima, M., Simpson, S.L., Li, J., Tscharke, B., Verhagen, R., Smith, W.J.M., Zaugg, J., Dierens, L., Hugenholtz, P., Thomas, K.V., Mueller, J.F.: First confirmed detection of SARSCoV-2 in untreated wastewater in Australia: a proof of concept for the wastewater surveillance of COVID-19 in the community. Sci. Total Environ. **728**, (2020). https://doi.org/10.1016/j.scitotenv.2020.138764
5. Ahmed, A., Hussein, S.E.: Leaf identification using radial basis function neural networks and SSA based support vector machine. PLoS ONE **15**(8), (2020). https://doi.org/10.1371/journal.pone.0237645
6. Alley, W.M., Alley R.: Pathogens. In: High and Dry: Meeting the Challenges of the World's Growing Dependence on Groundwater, pp. 195–202. New Haven: Yale University Press (2017)
7. American Society of Civil Engineers (ASCE): Failure to Act: The Economic Impact of Current Investment Trends in Water and Wastewater Treatment Infrastructure. Washington, DC (2011)
8. Amirian, E.S.: Potential fecal transmission of SARS-CoV-2: current evidence and implications for public health. Int. J. Infect. Dis. **95**, 363–370 (2020). https://doi.org/10.1016/j.ijid.2020.04.057
9. Ammendolia, J., Saturno, J., Brooks, A.L., Jacobs, S., Jambeck, J.R.: An emerging source of plastic pollution: environmental presence of plastic personal protective equipment (PPE) debris related to COVID-19 in a metropolitan city. Environ. Pollut. **269**, 116160 (2021)

10. Annalaura, C., Federigi, I., Dasheng, L., Julian, R.T., Marco, V.: Making waves: coronavirus detection, presence and persistence in the water environment: state of the art and knowledge needs for public health. Water Res. **179**, 115907 (2020). https://doi.org/10.1016/j.watres.2020. 115907
11. Asghar, H., Diop, O.M., Weldegebriel, G., Malik, F., Shetty, S., El Bassioni, L., Burns, C.C.: Environmental surveillance for polioviruses in the Global Polio eradication initiative. J. Infect. Dis. **210**(suppl_1), S294–S303 (2014)
12. Back, J.O., Rivett, M.O., Hinz, L.B., Mackay, N., Wanangwa, G.J., Phiri, O.L., Songola, C.E., Thomas, M.A.S., Kumwenda, S., Nhlema, M., Miller, A.V.M., Kalin, R.M.: Risk assessment to groundwater of pit latrine rural sanitation policy in developing country settings. Sci. Total Environ. **613–614**, 592–610 (2018). https://doi.org/10.1016/j.scitotenv.2017.09.071
13. Badawy, A.S., Rose, J.B., Gerba, C.P.: Comparative survival of enteric viruses and coliphage on sewage irrigated grass. J. Environ. Sci. Health Part A **25**(8), 937–952 (1990)
14. Bhallamudi, S.M., Kaviyarasan, R., Abilarasu, A., Philip, L.: Nexus between sanitation and groundwater quality: case study from a hard rock region in India. J. Water. Sanit. Hyg. Dev. **9**, 703–713 (2019). https://doi.org/10.2166/washdev.2019.002
15. Borchardt, M.A., Bertz, P.D., Susan K., Spencer, S.K., Battigelli, D.A.: Incidence of enteric viruses in groundwater from household wells in wisconsin. Appl. Environ. Microbiol. **69**, 1172–1180 (2003)
16. Carducci, A., Federigi, I., Liu, D., Thompson, J.R., Verani, M.: Making Waves: coronavirus detection, presence and persistence in the water environment: state of the art and knowledge needs for public health. Water Res. **179**, (2020). https://doi.org/10.1016/j.watres.2020.115907
17. Craun, G.F., Brunkard, J.M., Yoder, J.S., Roberts, V.A., Carpenter, J., Wade, T., Roy, S.L.: Causes of outbreaks associated with drinking water in the United States from 1971 to 2006. Clin. Microbiol. Rev. **23**(3), 507–528 (2010)
18. Datta, P.S., Deb, D.L., Tyagi, S.K.: Assessment of groundwater contamination from fertilizers in the Delhi area based on 18O, NO3 and K composition. J. Contam. Hydrol. **27**, 249–262 (1997)
19. DeSantis, C.E., Ma, J., Gaudet, M.M., et al.: Breast cancer statistics, 2019. CA Cancer J. Clin. **69**(6), 438–451 (2019)
20. Dimakakou, E., Johnston, H.J., Streftaris, G., Cherrie, J.W.: Exposure to environmental and occupational particulate air pollution as a potential contributor to neurodegeneration and diabetes: a systematic review of epidemiological research. Int. J. Environ. Res. Public Health **15**(8), 1704 (2018)
21. Dowd, S.E., Pillai, S.D., Wang, S., Corapcioglu, M.Y.: Delineating the specific influence of virus isoelectric point and size on virus adsorption and transport through sandy soils. Appl. Environ. Microbiol. **64**, 405–410 (1998)
22. El-Kenawy, E.S.M., Eid, M.M., Saber, M., Ibrahim, A.: MbGWO-SFS: Modified binary grey wolf optimizer based on stochastic fractal search for feature selection. IEEE Access **8**, 107635–107649 (2020). https://doi.org/10.1109/ACCESS.2020.3001151
23. El-Kenawy, E.S.M., Ibrahim, A., Mirjalili, S., Eid, M.M., Hussein, S.E.: Novel feature selection and voting classifier algorithms for COVID-19 classification in CT images. IEEE Access **8**, 179317–179335 (2020). https://doi.org/10.1109/ACCESS.2020.3028012
24. Espejo, W., Celis, J.E., Chiang, G., Bahamonde, P.: Environment and COVID-19: Pollutants, impacts, dissemination, management and recommendations for facing future epidemic threats. Sci. Total Environ. **747**, (2020)
25. Forster, P., Forster, L., Renfrew, C., Forster, M.: Phylogenetic network analysis of SARS-CoV-2 genomes. Proc. Natl. Acad. Sci. **117**(17), 9241–9243 (2020)
26. Forster, J., Saturno, J., Brooks, A.L., Jacobs, S., Jambeck, J.R.: An emerging source of plastic pollution: environmental presence of plastic personal protective equipment (PPE) debris related to COVID-19 in a metropolitan city. Environ. Pollut. 116160 (2020)
27. Fouad, M.M., El-Desouky, A.I., Al-Hajj, R., El-Kenawy, E.S.M.: Dynamic group-based cooperative optimization algorithm. IEEE Access **8**, 148378–148403 (2020). https://doi.org/10. 1109/ACCESS.2020.3015892

28. Gerba, C.P., Bitton, G.: Microbial pollutants: their survival and transport pattern to groundwater. In: Bitton, G., Gerba, C.P. (eds.) Groundwater pollution microbiology, pp. 65–68. John Wiley & Sons Inc, New York, N.Y (1984)

29. Gerba, C.P., Rose, J.B.: Viruses in source and drinking water. In: McFeters, G.A. (ed.) Drinking water microbiology: progress and recent developments, pp. 380–396. Springer-Verlag, New York, N.Y (1990)

30. Gov HK (Government of Hong Kong SAR): WHO Environmental Health Team reports on Amoy Gardens. https://www.info.gov.hk/gia/general/200305/16/0516114 (2003). Accessed 16 Sept 2020

31. Gupta, S., Parker, J., Smits, S., Underwood, J., Dolwani, S.: Persistent viral shedding of SARS-CoV-2 in faeces – a rapid review. Color. Dis. **22**, 611–620 (2020). https://doi.org/10.1111/codi.15138

32. Han, J., He, S.: Urban flooding events pose risks of virus spread during the novel coronavirus (COVID-19) pandemic. Sci. Total Environ. **755**, 142491 (2020). https://www.ngwa.org/publications-and-news/covid-19-news-and-resources/groundwater-wells-and-coronavirus

33. Hancock, C.M., Rose, J.B., Callahan, M.: Crypto and Giardia in US groundwater. JAWWA **90**(3), 58–61 (1998)

34. Haramoto, E., Malla, B., Thakali, O., Kitajima, M.: First environmental surveillance for the presence of SARS-CoV-2 RNA in wastewater and river water in Japan. Sci. Total Environ. **737**, (2020). https://doi.org/10.1016/j.scitotenv.2020.140405

35. Heller, L., Mota, C.R., Greco, D.B.: COVID-19 faecal-oral transmission: are we asking the right questions? Sci. Total Environ. **729**, (2020). https://doi.org/10.1016/j.scitotenv.2020.138919

36. Hill, M.K.: Understanding Environmental Pollution. Cambridge University Press (2020)

37. Hu, D., Zhang, Y., Shen, M.: Investigation on microplastic pollution of Dongting Lake and its affiliated rivers. Mar. Pollut. Bull. **160**, (2020)

38. Ibrahim, A., Mohammed, S., Ali, H.A., Hussein, S.E.: Breast cancer segmentation from thermal images based on chaotic Salp swarm algorithm. IEEE Access **8**, 122121–122134 (2020). https://doi.org/10.1109/ACCESS.2020.3007336

39. Ibrahim, A., Noshy, M., Ali, H.A., Badawy, M.: PAPSO: a power-aware VM placement technique based on particle swarm optimization. IEEE Access **8**, 81747–81764 (2020). https://doi.org/10.1109/ACCESS.2020.2990828

40. Ibrahim, A., Salem, M., Ali, H.A.: Block-based illumination-invariant representation for color images. Ain Shams Eng. J. **9**(4), 917–926 (2018). https://doi.org/10.1016/j.asej.2016.04.011

41. Ibrahim, A., El-kenawy, E.S.M.: Image segmentation methods based on superpixel techniques: a survey. J. Comput. Sci. Inf. Syst. **15** (2020a). Accessed 3 Oct 2020

42. Ibrahim, A., El-kenawy, E.S.M.: Applications and datasets for superpixel techniques: a survey. J. Comput. Sci. Infor. Syst. **15** (2020b), Accessed 3 Oct 2020

43. Ibrahim, A., Horiuchi, T., Tominaga, S., Hassanien, A.E.: Spectral reflectance images and applications. In: Image Feature Detectors and Descriptors, pp. 227–254. Springer, Cham (2016). https://doi.org/10.1007/978-3-319-28854-3_9

44. Ibrahim, A., Horiuchi, T., Tominaga, S., Hassanien, A.E.: Color Invariant Representation and Applications. In: Handbook of Research on Machine Learning Innovations and Trends, pp. 1041–1061. IGI Global (2017). https://doi.org/10.4018/978-1-5225-2229-4.ch046

45. Jeong, H.W., Kim, S.M., Kim, H.S., Kim, Y.I., Kim, J.H., Cho, J.Y., Kim, S., Kang, H., Kim, S.G., Park, S.J., Kim, E.H., Choi, Y.K.: Viable SARS-CoV-2 in various specimens from COVID-19 patients. Microbiol. Infect, Clin (2020). https://doi.org/10.1016/j.cmi.2020.07.020

46. Jin, Y., Flury, M.: Fate and transport of viruses in porous media. Adv. Agron. **77**, 39–102 (2002)

47. Kampf, G., Todt, D., Pfaender, S., Steinmann, E.: Persistence of coronaviruses on inanimate surfaces and their inactivation with biocidal agents. J. Hosp. Infect. **104**(3), 246e251 (2020). https://doi.org/10.1016/j.jhin.2020.01.022

48. Kitajima, M., Ahmed, W., Bibby, K., Carducci, A., Gerba, C.P., Hamilton, K.A., Haramoto, E., Rose, J.B.: SARS-CoV-2 in wastewater: state of the knowledge and research needs. Sci. Total Environ. 139076 (2020). https://doi.org/10.1016/j.scitotenv.2020.139076
49. Kumura, T., Suzuki, N., Takahashi, M., Tominaga, S., Morioka, S., Ivan, S.: Smart water management technology with intelligent sensing and ICT for the integrated water systems. Special Issue Solut. Soc. Creat. Safer More Secure Soc. 9(1), 103–106 (2015)
50. Lahrich, S., Laghrib, F., Farahi, A., Bakasse, M., Saqrane, S., El Mhammedi, M.A.: Review on the contamination of wastewater by COVID-19 virus: Impact and treatment. Sci. Total Environ. 751, (2021)
51. Lee, H.J., Kim, K.Y., Hamm, S.Y., Kim, M., Kim, H.K., Oh, J.E.: Occurrence and distribution of pharmaceutical and personal care products, artificial sweeteners, and pesticides in groundwater from an agricultural area in Korea. Sci. Total Environ. 659, 168–176 (2019)
52. Lescure, F., Bouadma, L., Nguyen, D., Parisey, M., Wicky, P., Behillil, S., Gaymard, A.: Clinical and virological data of the first cases of COVID-19 in Europe: a case series. Lancet Infect. Dis. 2, 1–10 (2020). https://doi.org/10.1016/S1473-3099(20)30200-0
53. Ling, Y., Xu, S., Lin, Y., Tian, D., Zhu, Z., Dai, F., Wu, F., Song, Z., Huang, W., Chen, J., Hu, B., Wang, S., Mao, E., Zhu, L., Zhang, W., Lu, H.: Persistence and clearance of viral RNA in 2019 novel coronavirus disease rehabilitation patients. Chin. Med. J. 133, E007 (2020). https://doi.org/10.1097/CM9.0000000000000774
54. Liu, Y., Chen, S., Zühlke, L., et al.: Global birth prevalence of congenital heart defects 1970–2017: updated systematic review and meta-analysis of 260 studies. Int. J. Epidemiol. 48(2), 455–463 (2019)
55. Lodder, W., de Roda Husman, A.M.: SARS-CoV-2 in wastewater: potential health risk, but also data source. lancet. Gastroenterol. Hepatol. 1253, 30087 (2020). https://doi.org/10.1016/S2468-1253(20)30087-X
56. Mao, K., Zhang, H., Yang, Z.: Can a paper-based device trace COVID-19 sources with wastewater-based epidemiology? Environ. Sci. Technol. 3733–3735 (2020). https://doi.org/10.1021/acs.est.0c01174
57. Marzouk, Y., Goyal, S.M., Gerba, C.P.: Prevalence of enteroviruses in ground water of Israel. Groundwater 17(5), 487–491 (1979)
58. Mavrikou, S., Georgia Moschopoulou, V.T., Kintzios, S.: Development of a portable, ultra-rapid and ultra-sensitive cell-based biosensor for the direct detection of the SARS- CoV-2 S1 spike protein antigen. Sensors 20, 3121 (2020). https://doi.org/10.3390/s20113121
59. Medema, G., Heijnen, L., Elsinga, G., Italiaander, R., Brouwer, A.: Presence of SARSCoronavirus-2 in sewage and correlation with reported COVID-19 prevalence in the early stage of the epidemic in the Netherlands. Environ. Sci. Technol. Lett. (2020). https://doi.org/10.1021/acs.estlett.0c00357
60. Miles, F.: Coronavirus wipes, masks a nightmare for storm drains, sewers. https://www.foxnews.com/us/coronavirus-wipes-masks-a-nightmare-for-storm-drainssewers (2020). Accessed 16 Sept 2020
61. Miller, S.A., Horvath, A., Monteiro, P.J.: Impacts of booming concrete production on water resources worldwide. Nature Sustain. 1(1), 69–76 (2018). https://doi.org/10.1016/j.jclepro.2017.04.040
62. Naddeo, V., Liu, H.: Editorial perspectives: 2019 novel coronavirus (SARS-CoV-2): what is its fate in urban water cycle and how can the water research community respond? Environ. Sci. Water Res. Technol. 6, 1213–1216 (2020). https://doi.org/10.1039/d0ew90015j
63. Nasser, A.A., Rashad, M.Z., Hussein, S.E.: A two-layer water demand prediction system in urban areas based on micro-services and LSTM neural networks. IEEE Access 8, 147647–147661 (2020). https://doi.org/10.1109/ACCESS.2020.3015655
64. Nghiem, L.D., Morgan, B., Donner, E., Short, M.D.: The COVID-19 pandemic: considerations for the waste and wastewater services sector. Case stud. Chem. Environ. Eng. 1, (2020). https://doi.org/10.1016/j.cscee.2020.100006
65. Omosa, I.B., Wang, H., Cheng, S., Li, F.: Sustainable tertiary wastewater treatment is required for water resources pollution control in Africa. Environ. Sci. Technol. 46, 7065–7066 (2012). https://doi.org/10.1021/es3022254

66. O'Brien, E.O., Xagoraraki, I.: A water-focused one-health approach for early detection and prevention of viral outbreaks. One Health **7**, (2020)

67. Pan, X., Chen, D., Xia, Y.: Viral load of SARS-CoV-2 in clinical samples. Lancet Infect. Dis. **20**, 411–412 (2020). https://doi.org/10.1016/S1473-3099(20)30113-4

68. Piasecki, A., Jurasz, J., Skowron, R.: Forecasting surface water level fluctuations of lake Serwy (Northeastern Poland) by artificial neural networks and multiple linear regression. J. Environ. Eng. Landscape Manag. **25**(4), 379–388 (2017)

69. Polo, D., Quintela-Baluja, M., Corbishley, A., Jones, D.L., Singer, A.C., Graham, D.W., Romalde, J.L.: Making waves: wastewater-based epidemiology for COVID-19–approaches and challenges for surveillance and prediction. Water Res. **186**, (2020)

70. Pooi, C.K., Ng, H.Y.: Review of low-cost point-of-use water treatment systems for developing communities. Clean Water 1 (2018). https://doi.org/10.1038/s41545-018-0011-0

71. Qu, G., Li, X., Hu, L., Jiang, G.: An imperative need for research on the role of environmental factors in transmission of novel coronavirus (COVID-19). Environ. Sci. Technol. **54**, 3730–3732 (2020). https://doi.org/10.1021/acs.est.0c01102

72. Randazzo, W., Truchado, P., Cuevas-Ferrando, E., Simón, P., Allende, A., Sánchez, G.: SARS-CoV-2 RNA in wastewater anticipated COVID-19 occurrence in a low prevalence area. Water Res. **181**, (2020). https://doi.org/10.1016/j.watres.2020.115942

73. Ribbers, K., Breuer, L., Düring, R.A.: Detection of artificial sweeteners and iodinated X-ray contrast media in wastewater via LC-MS/MS and their potential use as anthropogenic tracers in flowing waters. Chemosphere **218**, 189–196 (2019)

74. Rimoldi, S.G., Stefani, F., Gigantiello, A., Polesello, S., Comandatore, F., Mileto, D., Maresca, M., Longobardi, C., Mancon, A., Romeri, F., Pagani, C., Cappelli, F., Roscioli, C., Moja, L., Gismondo, M.R., Salerno, F.: Presence and infectivity of SARS-CoV-2 virus in wastewaters and rivers. Sci. Total Environ. **744**, (2020). https://doi.org/10.1016/j.scitotenv.2020.140911

75. Saad, A.K.: Environmental hydrogeologic impacts of groundwater withdrawal in the eastern Nile Delta region with emphasis on groundwater pollution potential. Ph.D. Thesis, Institute of Environmental Studies. Ain Shams Univ. Cairo, Egypt, p. 232 (1997)

76. Samia, K., Dhouha, A., Anis, C., Ammar, M., Rim, A., Abdelkrim, C.: Assessment of organic pollutants (PAH and PCB) in surface water: sediments and shallow groundwater of Grombalia watershed in northeast of Tunisia. Arab. J. Geosci. **11**(2), 34 (2018)

77. Shen, C.: A transdisciplinary review of deep learning research and its relevance for water resources scientists. Water Resour. Res. **54**(11), 8558–8593 (2018). https://doi.org/10.1029/2018wr022643

78. Sims, N., Kasprzyk-Hordern, B.: Future perspectives of wastewater-based epidemiology: monitoring infectious disease spread and resistance to the community level. Environ. Int. **139**(February), (2020). https://doi.org/10.1016/j.envint.2020.105689

79. Stefania, G.A., Rotiroti, M., Buerge, I.J., Zanotti, C., Nava, V., Leoni, B., Bonomi, T.: Identification of groundwater pollution sources in a landfill site using artificial sweeteners, multivariate analysis and transport modeling. Waste Manag. **95**, 116–128 (2019)

80. Sun, J., Zhu, A., Li, H., Zheng, K., Zhuang, Z., Chen, Z., Shi, Y., Zhang, Z., Chen, S.B., Liu, X., Dai, J., Li, X., Huang, S., Huang, X., Luo, L., Wen, L., Zhuo, J., Li, Y., Wang, Y., Zhang, L., Zhang, Y., Li, F., Feng, L., Chen, X., Zhong, N., Yang, Z., Huang, J., Zhao, J., Li, Y.M.: Isolation of infectious SARS-CoV-2 from urine of a COVID-19 patient. Emerg. Microb. Infect. **9**, 991–993 (2020). https://doi.org/10.1080/22221751.2020.1760144

81. Taha, A.A., El-Mahmoudi, A.S., El-Haddad, I.M.: Pollution sources and related environmental impacts in the new communities southeast Nile delta. Egypt. Emirates J. Engi. Res. **9**(1), 35–49 (2004)

82. Tang, A., Tong, Z., Wang, H., Dai, Y., Li, K., Liu, J., Wu, W., Yuan, C., Yu, M., Li, P., Yan, J.: Detection of novel coronavirus by RT-PCR in stool specimen from asymptomatic child, China. Emerg. Infect. Dis. J. 26 (2020). https://doi.org/10.3201/eid2606.200301

83. Tharanyaa, J., Jagadeesan, A., Lavanya, A.: Theft identification and automated water supply system using embedded technology. Inter. J. Adv. Res. Electr. Electron. Instrum. Eng. **2**(8), 3727–3733 (2013)

84. Thompson, L.A., Darwish, W.S.: Environmental chemical contaminants in food: review of a global problem. J. Toxicol. (2019)
85. Tierney, J.T., Sullivan, R., Larkin, E.P.: Persistence of poliovirus 1 in soil and on vegetables grown in soil previously flooded with inoculated sewage sludge or effluent. Appl. Environ. Microbiol. **33**(1), 109–113 (1977)
86. Tran, N.H., Hu, J., Li, J., Ong, S.L.: Suitability of artificial sweeteners as indicators of raw wastewater contamination in surface water and groundwater. Water Res. **48**, 443–456 (2014)
87. Tsatsakis, A., Petrakis, D., Nikolouzakis, T.K., Docea, A.O., Calina, D., Vinceti, M., Hernández, A.F.: COVID-19, an opportunity to reevaluate the correlation between long-term effects of anthropogenic pollutants on viral epidemic/pandemic events and prevalence. Food Chem. Toxicol. **141**, (2020). https://doi.org/10.1016/j.fct.2020.111418
88. Venugopal, A., Ganesan, H., Sudalaimuthu Raja, S.S., Govindasamy, V., Arunachalam, M., Narayanasamy, A., Sivaprakash, P., Rahman, P.K.S.M., Gopalakrishnan, A.V., Siama, Z., Vellingiri, B.: Novel wastewater surveillance strategy for early detection of coronavirus disease 2019 hotspots. Curr. Opin. Environ. Sci. Heal. **17**, 8–13 (2020). https://doi.org/10.1016/j.coesh.2020.05.003
89. WHO: Progress on sanitation and drinking-water: fast facts. https://www.who.int/water_sanitation_health/monitoring/jmp_fast_facts/en/ (2013). Accessed 8 May 2020
90. WHO: Modes of transmission of virus causing COVID-19: implications for IPC precaution recommendations. https://www.who.int/news-room/commentaries/detail/modes-of-transmission-of-virus-causing-covid-19-implications-for-ipc-precautionrecommendations (2020a). Accessed 8 May 2020
91. WHO: Coronavirus disease (COVID-19) advice for the public. https://www.who.int/emergencies/diseases/novel-coronavirus-2019/advice-for-public (2020b). Accessed 10 May 2020
92. Wang, J., Feng, H., Zhang, S., Ni, Z., Ni, L., Chen, Y., Zhuo, L., Zhong, Z., Qu, T.: SARSCoV-2 RNA detection of hospital isolation wards hygiene monitoring during the Coronavirus Disease 2019 outbreak in a Chinese hospital. Int. J. Infect. Dis. **94**, 103–106 (2020). https://doi.org/10.1016/j.ijid.2020.04.024
93. Wang, G., Lu, J., Tong, Y., Liu, Z., Zhou, H., Xiayihazi, N.: Occurrence and pollution characteristics of microplastics in surface water of the Manas River Basin China. Sci. Total Environ. **710**, (2020)
94. Wang, W., Xu, Y., Gao, R., Lu, R., Han, K., Wu, G., Tan, W.: Detection of SARSCoV-2 in different types of clinical specimens. J. Am. Med. Assoc. 3–4 (2020c). https://doi.org/10.1001/jama.2020.3786
95. Wright, D.B., Bosma, C.D., Lopez-Cantu, T.: US hydrologic design standards insufficient due to large increases in frequency of rainfall extremes. Geophys. Res. Lett. **46**, 8144–8153 (2019). https://doi.org/10.1029/2019GL083235
96. Xiao, F., Tang, M., Zheng, X., Liu, Y., Li, X., Shan, H.: Evidence for gastrointestinal infection of SARS-CoV-2. Gastroenterology **158**, 1831–1833.e3 (2020). https://doi.org/10.1053/j.gastro.2020.02.055
97. Xing, Y., Lu, Y., Dawson, R.W., Shi, Y., Zhang, H., Wang, T., Liu, W., Ren, H.: A spatial temporal assessment of pollution from PCBs in China. Chemosphere **60**, 731–739 (2005)
98. Xu, H.L., Ye, M., Li, J.M.: The water transfer effects on agricultural development in the lower Tarim River Xinjiang of China. Agric. Water Manag. **95**, 59–68 (2008)
99. Xu, K., Cai, H., Shen, Y., Ni, Q., Chen, Y., Hu, S., Li, J., Wang, H., Yu, L., Huang, H., Qiu, Y., Wei, G., Fang, Q., Zhou, J., Sheng, J., Liang, T., Li, L.: Management of COVID-19: the Zhejiang experience. J. Zhejiang Univ. Med. Sci. **49**, 147–157 (2020). https://doi.org/10.3785/j.issn.1008-9292.2020.02.02
100. Yeo, C., Kaushal, S., Yeo, D.: Enteric involvement of coronaviruses: is faecal−oral transmission of SARS-CoV-2 possible? Lancet Gastroenterol. Hepatol. **5**, 335–337 (2020). https://doi.org/10.1016/S2468-1253(20)30048-0
101. Yu, I.T.S., Li, Y., Wong, T.W., Tam, W., Chan, A.T., Lee, J.H.W., Leung, D.Y.C., Ho, T.: Evidence of airborne transmission of the severe acute respiratory syndrome virus. N. Engl. J. Med. **350**, 1731–1739 (2004). https://doi.org/10.1056/NEJMoa032867

102. Yu, M.-H., Tsunoda, H., Tsunoda, M.: Environmental Toxicology: Biological and Health Effects of Pollutants. CRC press (2011)

103. Zhang, C., Wang, S., Sun, D., Pan, Z., Zhou, A., Xie, S., Zou, J.: Microplastic pollution in surface water from east coastal areas of Guangdong, South China and preliminary study on microplastics biomonitoring using two marine fish. Chemosphere 127202 (2020a)

104. Zhang, D., Fraser, M.A., Huang, W., Ge, C., Wang, Y., Zhang, C., Guo, P.: Microplastic pollution in water, sediment, and specific tissues of crayfish (Procambarus clarkii) within two different breeding modes in Jianli, Hubei province, China. Environ. Pollut. 115939 (2020b)

105. Zhang, Y., Chen, C., Zhu, S., Shu, C., Wang, D., Song, J., Song, Y., Zhen, W., Feng, Z., Wu, G., Xu, J., Xu, W.: Isolation of 2019-nCoV from a stool specimen of a laboratory confirmed case of the Coronavirus Disease 2019 (COVID-19). China CDC Weekly **2**(8), 123–124 (2020c). https://doi.org/10.46234/ccdcw2020.033

106. Zhu, Z., Khattab, F., Elassiouti, I., Abdel Azim, R., Keith, J., Mahdy, E., Cardinalli, R., Galal Saad, S., Mohiedin, M.: Intermediate drainage reuse in Bahr Bagar drain basin, United States Agency for International Development/Egypt. Task Order No. 807 Contract No. PCE-I-00-96-00002-00, p. 74 (2020c)

Web Resources

Daily reports of the Egyptian Ministry of Health and Population, 2020. COVID-19 infection cases in Sharkia Governorate. https://bit.ly/3da6pG0

Environmental Challenges and the Impact of COVID-19 on Healthcare Sector: The Adoption of Intelligent Solutions

Yasmine S. Moemen and Ibrahim El-Tantawy El-Sayed

Abstract The current epidemic disease is a Severe Acute Respiratory Syndrome CoronaVirus 2 (SARS-CoV-2) or CoronaVirus Disease 2019 (COVID-19) has produced many morbidities and mortality worldwide and conquer all countries globally, spreading till this moment and cause a health hazard for the coming years. Recent technologies such as robotics, sensor networks and autonomous vehicles (AV) have been identified as a new strategy that would make an essential. To fight against COVID-19. This technology provides remote safety for healthcare works and professionals, patients and can provide faster reporting, virtual visits, results and more. The environment is an essential constituent of both human and animal health. The COVID-19 epidemic has massive influences on most human actions, besides the economy and health care organizations. Lock-downs, quarantines and border closings had reduced air pollution through reduced travel and production. The environmental profits are temporary, but this can positively affect the environment and technology use like teleconferencing and telehealth. This plague is a chance for the government to reconsider using more technology in the health sector, like more supported computational studies and artificial intelligence to put scenarios to confine the virus. This chapter concludes the recent knowledge that provides the best community healthcare requirements, enhances care performance develops treatments and vaccines.

Keywords COVID-19 · Environmental

Y. S. Moemen (✉)
Clinical Pathology Department, National Liver Institute, Menoufia University, Menoufia, Egypt
e-mail: yasmine_moemen@liver.menofia.edu.eg

I. E.-T. El-Sayed
Chemistry Department, Faculty of Science, Menoufia University, Menoufia, Egypt

A. E. Hassanien et al. (eds.), *The Global Environmental Effects During and Beyond COVID-19*, Studies in Systems, Decision and Control 369,
https://doi.org/10.1007/978-3-030-72933-2_3

1 Introduction

The current outbreak is a Severe Acute Respiratory Syndrome CoronaVirus 2 (SARS-CoV-2), it is similar to Severe Acute Respiratory Syndrome CoronaVirus (SARS-CoV) by 79.6%; which has been appeared in China during 2002–2003 [1] and it declared as 'a worldwide plague' by WHO on March 11, 2020. SARS-CoV-2 has been invaded 212 nations and areas worldwide. It is the second global transmissions after the Spanish Flu pandemic triggered by the H1N1 virus. On 29 October 2020, about 44,710,816 patients have been detected with 1,178,021 mortalities and 32,680,963 recovered patients. Currently active infected patients are 10,851,832 while 10,770,771(99%) are in weak illness and 81,061 (1%) patients are in dangerous illness [2]. To the best of my knowledge, there are seven coronaviruses (CoVs) with the strains 229E (α-CoV), HKU1 (β-CoV), OC43 (β-CoV) and NL63 (α-CoV) which produced weak symptoms of the upper respiratory tract in individual [3]. Furthermore, another strains were already exist as SARS-CoV (2002–2003), the Middle East respiratory syndrome happening in 2012 (MERS-CoV), and the novel defined SARS-CoV-2 related to β-CoV has severe symptoms that may lead to death [4].

Healthcare structures have put great trials to cure the infected cases and test the public for COVID-19 infection and to confront the situation. Countries are doing their role to extenuate the epidemic and even to meet the health care system requirements. Moreover, there is unfortunately neither a cure medicine nor a prevention vaccine for this novel coronavirus. Most countries have attempted various forms of treatment and the use of specific effective drugs against various illness (without adequate confirmation). The WHO's protective measures are washing hands frequently or hand rubbing using alcohol besides keep social distance, exercising, breathing fresh air and hygiene. Wearing a mask protects a person against infection besides learning when and how to use the mask which is essential. Accordingly, the only way to avoid the infection is by remaining at home, preventing social interactions, adopting social distancing and isolating us from infected persons or persons which submitted to isolation. It also takes collaborative efforts from community institutions, like healthcare structure, authority and most specifically, the community to avoid the outbreak. Which, they can only be controlled by a restricted degree. The exhaustive use of prospective knowledge, good health care medication and powerful management will strengthen the defense line against COVID-19.

Most individuals infected with SARS-CoV-2 symptoms will have mild to moderate respiratory disease that might recover without physician consultation. Individuals with chronic diseases like cardiovascular, diabetes, chronic respiratory disease and cancer might develop acute respiratory symptoms. The prevention of COVID-19 transmission should include keeping social distance among people, cleaning hands or using sanitizers and not touching the face. It transmits through droplets of saliva from the infected individual coughs or sneezes [5]. Presently, no available inoculation or cure for COVID-19. However, many current clinical trials are evaluating potential drugs [6, 7, 8].

With many changes forced upon our daily lives by the COVID-19 pandemic, the transportation industries and logistics supplies are on the lookout for how consumer behavior changes may affect the adoption of Audio Video (AV) technologies across all facets of the economy and daily life. Autonomous AVs, also known as self-driving cars, can sense their environment and move safely with almost no human interaction. AV technology's very promising to enhance community safe and offer common benefits, like reducing the number of morbidities or mortalities caused by road crashes, improving traffic and making transport accessible to older adults or disabled people. There are also many applications of AVs technologies for facing a pandemic such as:

- Autonomous logistic vehicle for contact-less medical supplies, food and packages transportation during the outbreak.
- AVs to sanitize regions and formed partnerships to accelerate AV technology development to lessen the virus's spread.
- Robotics services were deployed to provide an alternative transportation system for people.

The remain of this chapter is arranged as the following structure. Section 2 presents the basics and background. Section 3 reviews previous methods and experiments. Section 4 introduces the analysis and discusses these studies. Section 5 concludes this chapter.

2 Conceptual Background

2.1 Healthcare Sector in the Age of COVID-19

At the age of COVID-19, health care technologies have been progressed to cover the clinical needs, originating from COVID-19 confirming to virus confining, where various studies [6, 7, 9, 10, 11] have been executed and trialed. For more illustration, the confirmative measures for COVID-19 have been passed in diverse stages, from rapid tests with less precision to high precision testing in case of time adjustment. Commonly, Polymerase Chain Reaction (PCR) is the approved test for the analysis of COVID-19, which is the most accurate technique which measures the RNA of the virus. There are other testing procedures like serologic testing and antigen testing [12]. Testing many people is an urgent demand to evaluate the overall transmission of the SARS-CoV-2 virus. The need for testing tools is increasing. The procedures used for testing vary from nation to other. For crowded nations, antibody examinations effectively monitor the people's immune nature and describe the virus transmission rate. Furthermore, they can yield results in the short-term and fast antibody testing which would take about 15–30 min to produce results [13]. Bosch improved an advanced fast test to yield the results in a short time about 2 and half an hour as he claims that it is one of the world's entirely programmed molecular diagnostic tests

[12]. A San-Francisco group called Cough for the Cure, researchers and engineers used a model to classify COVID-19 the sounds of the patient's cough [12]. It is effective, but it is not approved because it is not known if COVID-19 cough is unique or not, although it is novel. The usage of telehealth facilities increasing very rapidly during the epidemic period.

2.2   Intelligent Solutions

In this regard, recent technology is a critical tool that can provide the following advantages:

1.   Protection of health patients, workers and professionals
2.   Enabling the work and medicine remotely
3.   It is increasing the access and processing of healthcare data and medical records.
4.   We are providing supplies to hospitals as with autonomous vehicles in the lockdown times.
5.   They are providing care, safety to populations and communities.

Artificial Intelligence is used during the COVID-19 epidemic for the detection and diagnosis of COVID-19, examining the efficacy of cure, virus spread prediction, cure and vaccine improvement.

2.2.1   Robotics

Robots are sophisticated systems that have been used in the last decades in healthcare systems. Recently, the COVID-19 crisis has jet-fueled the adoption of robots as an assistive tool in fighting COVID-19. Robots have been used for delivering supplies and medications to patients in hospitals and infected areas, where these activities need human interaction. The implementation of robots in healthcare sector like a means to enhance the performance of healthcare sector during pandemics such as COVID-19. Robots' services can eliminate human interaction and allow healthcare workers to be organized to the highest priority area necessary [14].

The robots have particular responsibilities like taking temperatures, communicating with infected patients and making beds. Robots can also understand, speak general concepts and they can move in smooth, fluid, human-like ways. Moreover, the COVID-19 pandemic also witnessed an increase in room cleaning and UV light robots. Finally, the use of robotic surgery has increased in urology and gynecology in the last two decades. Recent studies suggest minimizing the number of staff who are participating in operating rooms during the COVID-19 pandemic

2.2.2 Autonomous Vehicles

AV has attracted attention over the last decade and different vendors developed prototype models. Although the commercial implementation of autonomous vehicles remains a major challenge. At a fundamental level, AV contain a variety of sensors and actuators which produce much real-time data which will produce decisions. The AV design must also consider the quality, speed, volume, diversity and real-time of data nature. It is important to note that various automobile manufacturers use on-board sensor and actuator technology in various optimized applications. However, the main requirement of operating independently is at the heart of the AV design. In other words, the AV requires capabilities that allow it to expect, define and move according to a certain strategy. The scenarios for enhancing comfort and life quality are limitless AV would turn the entire transport infrastructure into a new model of linked smart ecosystems. AVs also change the future of logistics delivery, health care assistance and many other scenarios.

The AV have a role during the COVID-19 pandemic like logistic delivery, transportation and disinfection. Because of the quick outbreak of COVID-19, many day-to-day activities, such as the logistics of transporting medical supplies, food and packages in daily lives, inevitably involve contact between people. In many countries, AV are used to prevent person-to-person communication as an appropriate solution to control and prevent infections. AV also provides an alternative to traditional transportation with many benefits like safety, cost, avoidance of accidents and communications between people. In China, Due to the COVID-19 lockdown transportation during the Chinese New Year period at the end of January, the level of transport decreased by 50% in 2019. Public health panic has improved confidence in autonomous vehicles (self-driving cars). Wuhan's citizens have increasingly accepted self-driving vehicles to provide deliveries and transport people to certain places safely in countless situations involving small traveler packages. AVs are also used to disinfect areas with no human contact to mitigate the virus's spread. Intel presents Autonomous Virus-Killing Robots using UV lights to help overburdened hospitals in the fight against COVID-19. Besides, AV and robotics are used in patient care. In China, the robotics company CloudMinds deploy robots in Wuhan Hongshan Stadium after being converted into a smart field hospital. These robots were used to sanitize, to evaluate patient temperatures, to provide medicine and food. In Italy, robots have been used to reduce direct contact between patients and medical staff while looking after their patients [15].

2.2.3 Internet of Things

Internet of Things (IoT) is utilized for different applications to satisfy the significant requirements to mitigate the COVID-19 epidemic results. It can predict the future status and to correct the collected. IoT solutions are used to manage this pandemic properly. IoT services can be used to better monitor glucometer, blood pressure, heart rate, and other vital parameters especially monitor the decrease Oxygen level which is one of the common COVID-19 complications. Moreover, about 10–15% of COVID-19 patients need to put them in the Intensive Care Unit (ICU). IoT-based ICU equipped for immediate control of multiple health-related problems and alert generation procedures, including information distribution for specific vital circumstances. IoT helps track older people's health conditions, especially they are the most affected by COVID-19 [16].

These technologies' important applications against COVID-19 are to follow the real-time site of infected and suspected people for a seamless process of care with no interruption. Moreover, IoT improves the treatment workflow with effective results and assists in the decision-making process during complicated cases. Another Significant IoT application against COVID-19 is a real-time screening of the suspected people. Moreover, During the COVID-19 epidemic, IoT-based platforms for drug supply and pharmaceutical management practices have more relied on, covering four aspects: drug supply management, drug use management off-label, pharmaceutical care and individuals. Moreover, during the pandemic, many people were forced to work through the home, so the IoT provided the tools necessary to monitor COVID-19 patients.

2.2.4 5G Based Healthcare for COVID-19

COVID-19 epidemic triggered a huge effect on healthcare, public life, and financial system on a worldwide level. Technology enables abundant and accessible digital health facilities in epidemic situations versus the "re-emergence" of COVID-19 illness in a post-epidemic era. Consequently, 5G systems and 5G-enabled e-health solutions are vital. Facilities of 5G technologies can be successfully used to focus on the tasks related to COVID-19 at the moment and in the post-COVID-19 era. Current healthcare facilities should be designed to match the requirements of the COVID-19 era while improving new solutions to define the particular topics which come from the epidemic [17]. Solutions improved using 5G technologies serve various health-related use cases such as telehealth, supply chain management, self-isolation, contact tracing and quick health services deployments.

Internet of Things devices can be connected to 5G network successfully as it used to observe self-isolation compliance. As an alternative of using general mobile device data, the patients can be connected with low-power wearable devices that transmit data via BLE technology. Those sensory data can be renewed to the cloud via the 5G network and the authority can watch the patient behavior.

2.3 Drug Discovery

The current situation and existing morbid physiology postulates that SARS-CoV-2 is extremely infectious and spreadable than its ancestor, which targets different systems like respiration, the gastrointestinal and central nervous system besides various organs like the liver, heart, and kidney that leading to many organs deterioration [18]. SARS-CoV-2 has spike S glycoprotein, which has similar genes with 72% of human SARS and it is exclusive for furin-like cleavage site, which does not exist in any SARS-like CoVs [19]. Another work [20] revealed that SARS-CoV-2 binds by 10–20 times to the ACE2 receptor than SARS-CoV. This is because SARS-CoV-2 is unique in molecular diversity. It is higher by 1–3 times than SARS-CoV; which, may develop more mortality and morbidity [21]. So, the inhibition of the viral S protein ACE2 is significant approach for designing potential drugs for COVID-19 [22]. The SARS-CoV-2 replicates several steps targeting different proteins, so inhibition of such proteins will destroy the virus. The replication steps and involved proteins are [23], genomic RNA translation, proteolysis of translated RNA by proteinase, replication of genomic RNA using $3'$-to-$5'$ exonuclease RNA dependent RNA polymerase (RdRp), endoRNAse, helicase, $2'$-O-ribose methyltransferase and virus Assembly.

2.4 Drug Repositioning

Drug repositioning is a new treatment strategy that significantly reduces hazards associated with drug development and its costs. Repositioning is, besides its original indications, described as a new use of a drug. It is known as an alternative for rapidly detecting new therapeutic agents. While the number of FDA-approved medications has been declining steadily over the past three decades, product repurposing will promote new medicines.so, repurposing approved or fast-tracked clinical-stage investigational drugs into clinical trials can secure the much needed immediate relief [24].

In this emergency, the time lag between recognizing a potentially beneficial medication and treating the patient is shortened due to accessibility, high levels of protection, tolerability, pharmacokinetics, pharmacodynamics and clinical reports on medications currently used. Therefore, confronted with preliminary clinical effectiveness findings or a clear pharmacological basis, current medicines for new medical use in patients of humans may be evaluated immediately. As it is a major benefit of old medications, humans understand their toxicity and pharmacokinetic profile. Since it has become clear that the medicines approved for one disease could also be used for other indications, The FDA-approved drugs are attractive in terms of the time and the cost of drug production as a source of discovery of new drugs.

3   Review of Previous Methods and Experiments

3.1   Experimentation and Clinical Trials Approach

Currently, there is no available vaccine or appropriate drugs available to combat COVID-19. Most patients were treated based on symptoms if mild or critical (oxygen therapy with ventilator support). A group of available marketed drugs as hydroxy-chloroquine (HCQ) [6, 9], chloroquine (CLQ) [9], blend of HCQ and azithromycin [7], remdesivir [10], lopinavir [11] and ritonavir [11] are used for SARS-COV-2 treatment, Fig. 1; their clinical trials performed by various pharmaceutical companies. A set of novel vaccines still in clinical trial evaluations like reported [25] mRNA-1273, Ad5-nCoV [26] and ChAdOx1 nCoV19 [27] besides the existing Bacillus Calmette–Guerin (BCG) vaccine [28, 29] to evaluate its efficacy on COVID-19.

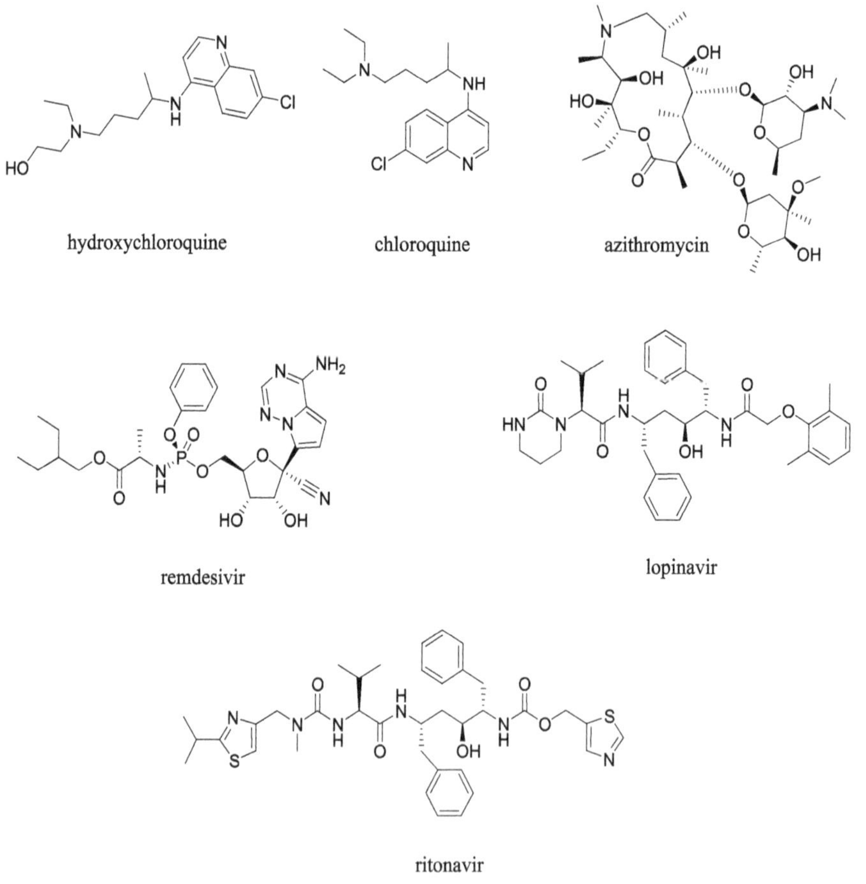

Fig. 1   Available marketed drugs for SARS-CoV-2 treatment

Many computational studies are performed on SARS-CoV-2, to discover several in silico methods and artificial intelligence approaches [30, 31, 32, 33, 34, 35] to detect the potential target and effective therapeutic molecular structure; which, are suitable for the drug discovery process. The RCSB protein data bank (PDB) (www.rcsb.org) has already deposited about 110 protein crystal structures related to SARS-CoV-2 and COVID-19 for permitting to understand significant binding sites of proteins that can be discovered in rational design of small molecular structures. The present work demonstrates the most revised and the drug nominees through an experiment used to heal patients globally. Besides, their potential targets and their mechanism of actions are not clear in most situations. Previous comprehensive review [36], reported computational studies describing data associated with in silico approaches, using protein crystal structure and series of potential inhibitors in the virtual screening (VS) approaches using docking, molecular dynamics (MD) and homology modeling. More comprehensive data related to SARS-CoV-2 spread, control of disease, protein structures, disease variety, diagnosis and testing are discussed in details in many literatures and reviews [1, 2, 4, 9–11].

4 Analysis and Discussion of the Results of the Previous Work

4.1 The Impact of COVID-19 on Social and Economic Life

The government should provide more improved decision-making procedures, handle community problems provide enough supplies for the healthcare system and prepare a health sector plan through the pandemic time. The actions commonly seen in authority are administrated organizations to cooperate, focusing on COVID problems and informing people about COVID-19 risk of infection. These social organizations can regulate the range to which technology is used for testing with novel procedures for diagnosis. After COVID-19, people's privacy is at risk to follow the virus transmitters in society and these actions are applied for saving the people's health. Governments are also starting arrangements, assisting scientific plans and actions that will help communicate with the public concerning health disasters [37]. Concerning the community, collaborating with the authority to confine the virus, individuals might have been unhappy from the frequent contact with the same surroundings throughout the isolation period. The flowing industries, cellular technology, broadband connections cannot follow quarantine rules. Working at home approaches and *E*-learning give society a chance to function as normal through the isolation period. From healthcare providers to the community, the entire system depends on the people's behaviors and collaboration.

A small number of helpful COVID-19 features, where the world is getting cleaned regularly. Through the quarantine period, air and pollution are reduced, besides, the whole environment is improved. Streams and seashores were arranged and cleaned due to quarantine requirements [38]. The release of greenhouse gases was low after World War II due to quarantine effects [38, 39]. It was been reported that the air was getting better where 55% particulate matter reduced in India through a primary lockdown period of 21 days [39]. Zambrano-Monserrate et al. [38] has started an environmental literature article of 367 cities of China. They reported that due to quarantine and lockdowns, NO2 was reduced by 22.8 and 12.9 $\mu g/m^3$ in Wuhan and China, respectively. PM 2.5 fell by 1.4 $\mu g/m^3$ in Wuhan but decreased by 18.9 $\mu g/m^3$ in 367 cities [38]. Sharma et al. [39] has reported that the concentrations of six criteria pollutants, PM10, PM2.5, CO, NO2, ozone and SO2, during mid-March to mid-April 2017–2020 in 22 cities covering different regions of India [39]. Overall, around 43, 31, 10 and 18% decreases in PM2.5, PM10, CO, and NO2 in India were observed during lockdown period compared to previous years. At the same time, there were 17% increase in O3 and negligible changes in SO2. The air quality index was reduced by 44, 33, 29, 15, and 32% in north, south, east, central and western India, respectively [39]. The other positive aspect learned from COVID-19 was that every country is trying its best to uplift the healthcare services.

In the year 2016, the World Health Organization (WHO), reported that 24% of worldwide mortality and 28% of these cases in children beneath five are due to environmental alternation issues. 68% of mortalities and 51% of diseases burden can be assessed with rational danger evaluation approaches, the calculations of other environmental exposures were finished by extra epidemiologic evaluations and skillful judgment. Ischemic heart disease, chronic respiratory diseases, cancers, and accidental injuries head happen to Individuals in low- and middle-income nations tolerate the extreme disease load as listed in supplementary Tables in 2016 and updated to the year 2019 [40] and as shown in Fig. 2.

This data table is an update of the publication "Preventing disease through healthy environments", pp. 107 (https://www.who.int/quantifying_ehimpacts/publications/preventing-disease/en/), WHO, Geneva, 2016. Data for a population of age 15 and above, Data for a population of age 25 and above, Data for children under the age of five years, Fraction of measles induced by malnutrition-related to the environment, LMIC: low- and middle-income countries, HIC:high-income countries, HIV/AIDS: human immunodeficiency virus infection/acquired immune deficiency syndrome.

5   Concluding Remarks

Several emerging technologies were introduced to tackle the unprecedented crisis of the new COVID-19 (coronavirus). As a result of the COVID-19 pandemic, everything will be transformed into a digital entity with some emerging technology. In this chapter, the conceptual foundation of the highlighted technologies for fighting the crisis of the COVID-19 pandemic, such as robots, IoT, 5G technology, autonomous vehicles, is outlined to assist patients and healthcare workers. The newly faced crisis of the COVID-19 pandemic needs a complex solution that not just a fusion of several

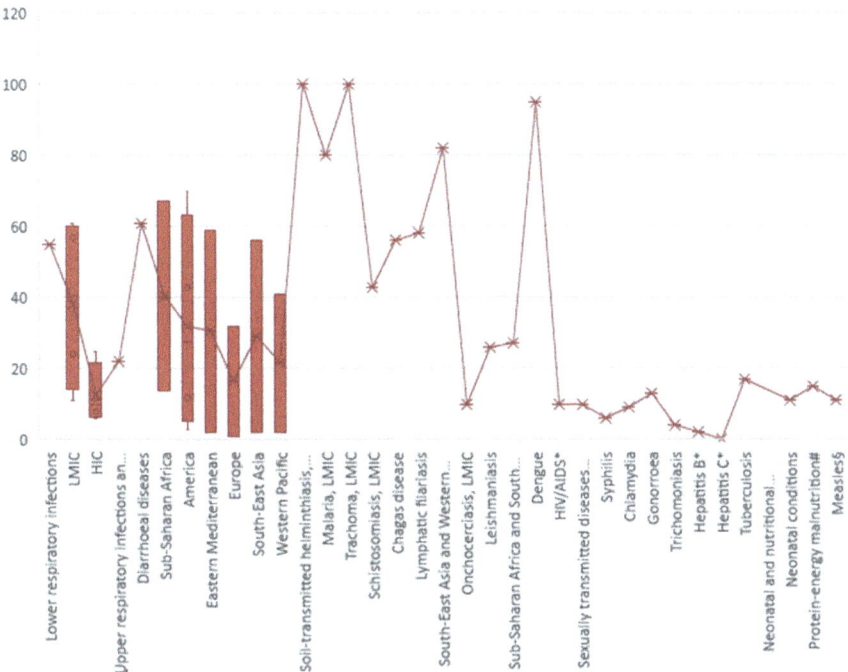

Fig. 2 Population attributable fractions for the environment (of DALYs), by disease and region

technologies such as cloud computing, big data analytics and IoT but also includes robotics. Drug repurposing as an approach based on machine learning techniques is presented as well.

References

1. Wu, F., et al.: A new coronavirus associated with human respiratory disease in China. Nature **579**(7798), 265–269 (2020)
2. Zhou, P., et al.: A pneumonia outbreak associated with a new coronavirus of probable bat origin. Nature **579**(7798), 270–273 (2020)
3. World Health Organization: COVID-19 Coronavirus Pandemic. https://www.worldometers. info/coronavirus/ (2020). Accessed 01 Oct 2020
4. Zeng, Z.-Q., et al.: Epidemiology and clinical characteristics of human coronaviruses OC43, 229E, NL63, and HKU1: a study of hospitalized children with acute respiratory tract infection in Guangzhou, China. Eur. J. Clin. Microbiol. Infect. Dis. **37**(2), 363–369 (2018)
5. World Health Organization: WHO Coronavirus Disease (COVID-19). https://www.worldomet ers.info/coronavirus/#countries (2020). Accessed 22 Oct 2020
6. Yao, X., et al.: In vitro antiviral activity and projection of optimized dosing design of hydroxychloroquine for the treatment of severe acute respiratory syndrome coronavirus 2 (SARS-CoV-2). Clin. Infect. Dis. (2020)

7. Gautret, P., et al.: Clinical and microbiological effect of a combination of hydroxychloro-quine and azithromycin in 80 COVID-19 patients with at least a six-day follow up: A pilot observational study. Travel Med. Infect. Dis. 101663 (2020)
8. Elavarasan, R.M., Pugazhendhi, R.: Restructured society and environment: a review on poten-tial technological strategies to control the COVID-19 pandemic. Sci. Total Environ. 138858 (2020)
9. Fantini, J., Di Scala, C., Chahinian, H., Yahi, N.: Structural and molecular modeling studies reveal a new mechanism of action of chloroquine and hydroxychloroquine against SARS-CoV-2 infection. Int. J. Antimicrob. Agents 105960 (2020)
10. Wang, M., et al.: Remdesivir and chloroquine effectively inhibit the recently emerged novel coronavirus (2019-nCoV) in vitro. Cell Res. **30**(3), 269–271 (2020)
11. Cao, B., et al.: A trial of lopinavir–ritonavir in adults hospitalized with severe Covid-19. N. Engl. J. Med. (2020)
12. Verdict Medical devices: Screening for Covid-19 (2020)
13. India Today: Rapid antibody tests (2020)
14. Health IT.gov. What Is Telehealth? How Is Telehealth Different From Telemedicine? Last reviewed October 17. https://www.healthit.gov/faq/what-telehealth-howtelehealth-different-telemedicine (2019). Accessed 02 June 2020
15. Zeng, Z., Chen, P.-J., Lew, A.A.: From high-touch to high-tech: COVID-19 drives robotics adoption. Tour. Geogr. 1–11 (2020)
16. Allam, Z., Jones, D.S.: On the coronavirus (COVID-19) outbreak and the smart city network: universal data sharing standards coupled with artificial intelligence (AI) to benefit urban health monitoring and management. Healthcare **8**(1), 46 (2020)
17. Elmousalami, H.H., Hassanien, A.E.: Day level forecasting for Coronavirus Disease (COVID-19) spread: analysis, modeling and recommendations. http://arxiv.org/abs/2003.07778 (2020)
18. Zhu, N., et al.: A novel coronavirus from patients with pneumonia in China, 2019. N. Engl. J. Med. (2020)
19. Li, X., Zai, J., Wang, X., Li, Y.: Potential of large 'first generation' human-to-human transmission of 2019-nCoV. J. Med. Virol. **92**(4), 448–454 (2020)
20. Coutard, B., Valle, C., de Lamballerie, X., Canard, B., Seidah, N.G., Decroly, E.: The spike glycoprotein of the new coronavirus 2019-nCoV contains a furin-like cleavage site absent in CoV of the same clade. Antiviral Res. **176**, (2020)
21. Song, W., Gui, M., Wang, X., Xiang, Y.: Cryo-EM structure of the SARS coronavirus spike glycoprotein in complex with its host cell receptor ACE2. PLoS Pathog. **14**(8), (2018)
22. Liu, C., et al.: Research and development on therapeutic agents and vaccines for COVID-19 and related human coronavirus diseases. ACS Publications (2020)
23. Fung, T.S., Liu, D.X.: Human coronavirus: host-pathogen interaction. Annu. Rev. Microbiol. **73**, 529–557 (2019)
24. Avchaciov, K., Burmistrova, O., Fedichev, P.O.: AI for the repurposing of approved or investigational drugs against COVID-19. Res. Gate. **10** (2020)
25. A Study to Evaluate Efficacy, Safety, and Immunogenicity of mRNA-1273 Vaccine in Adults Aged 18 Years and Older to Prevent COVID-19. ClinicalTrials.gov (2020)
26. Phase I Clinical Trial of a COVID-19 Vaccine in 18–60 Healthy Adults (CTCOVID-19). ClinicalTrials.gov (2020)
27. A Study of a Candidate COVID-19 Vaccine (COV001) ClinicalTrials.gov (2020)
28. de Vrieze, J.: Can a century-old TB vaccine steel the immune system against the new coronavirus?. Science (2020)
29. BCG Vaccination to Protect Healthcare Workers Against COVID-19 (BRACE). Clinical-Trials.gov
30. Huang, X., Pearce, R., Zhang, Y.: Computational design of peptides to block binding of the SARS-CoV-2 spike protein to human ACE2. bioRxiv (2020)
31. Smith, M., Smith, J.C.: Repurposing therapeutics for COVID-19: supercomputer-based docking to the SARS-CoV-2 viral spike protein and viral spike protein-human ACE2 interface (2020)

32. Ton, A., Gentile, F., Hsing, M., Ban, F., Cherkasov, A.: Rapid identification of potential inhibitors of SARS CoV 2 main protease by deep docking of 1.3 billion compounds. Mol. Inform. (2020)
33. Zhang, L., et al.: Crystal structure of SARS-CoV-2 main protease provides a basis for design of improved α-ketoamide inhibitors. Science **80368**(6489), 409–412 (2020)
34. Zhou, Y., Hou, Y., Shen, J., Huang, Y., Martin, W., Cheng, F.: Network-based drug repurposing for novel coronavirus 2019-nCoV/SARS-CoV-2. Cell Discov. **6**(1), 1–18 (2020)
35. Grifoni, A., Sidney, J., Zhang, Y., Scheuermann, R.H., Peters, B., Sette, A.: A sequence homology and bioinformatic approach can predict candidate targets for immune responses to SARS-CoV-2. Cell Host Microbe (2020)
36. Ojha, P.K., Kar, S., Krishna, J.G., Roy, K., Leszczynski, J.: Therapeutics for COVID-19: from computation to practices—where we are, where we are heading to. Mol. Divers. 1–35 (2020)
37. World Economic Forum: Government and companies response to COVID-19 (2020)
38. Zambrano-Monserrate, M.A., Ruano, M.A., Sanchez-Alcalde, L.: Indirect effects of COVID-19 on the environment. Sci. Total Environ. 138813 (2020)
39. Sharma, S., Zhang, M., Gao, J., Zhang, H., Kota, S.H.: Effect of restricted emissions during COVID-19 on air quality in India. Sci. Total Environ. **728**, (2020)
40. Updated 2016 data tables for 'Preventing disease through healthy environments'. Geneva: World Health Organization, 2019 (WHO reference number). Licence: CC BY-NC-SA 3.0 IGO. https://creativecommons.org/licenses/by-nc-sa/3.0/igo (2019)

COVID-19 Health Waste Management in Taiwan

Kai-Chun Chu and **Kuo-Chi Chang**

Abstract 2020 has brought a severe impact on the world. The COVID-19 epidemic has infected more than 30 million people worldwide, and more than a million people have died. International economic activities have also suffered unprecedented harm. Thereforiniiie, relevant personnel's response to the COVID-19 epidemic is achieved at all stages to prevent COVID-19 injury entirely. Medical institutions, centralized quarantine sites, home quarantine personnel, and all the people involved can adequately handle the waste generated by the COVID-19 epidemic. Therefore, this chapter has formulated management measures and operating principles which mainly constitute the following four sections includes (1) Introduction to health waste management; (2) Waste classification and cleaning methods in COVID-19 medical institutions; (3) Centralized quarantine station and home isolation waste cleaning method; (4) COVID-19 infection control and risk assessment measures in Taiwanese hospitals (5) Feasible application of the smart system in health waste management.

Keywords COVID-19 · Health wastesmanagement smart system · Medical institutions · COVID-19 infection control · Risk assessment measures

K.-C. Chu
School of Management, Fujian University of Technology, Fuzhou, China

Department of Business Administration Group of Strategic Management from,
National Central University, Taoyuan, Taiwan

Nurse Member of Taoyuan City Registered Professional Nurse Association, Taoyuan, Taiwan

K.-C. Chang (✉)
Fujian Provincial Key Laboratory of Big Data Mining and Applications, Fujian University of Technology, Fuzhou, China

Yu Da University of Science and Technology, Hsinchu, Taiwan

Fujian University of Technology, Fuzhou, China

College of Mechanical & Electrical Engineering, National Taipei University of Technology, Taipei, Taiwan

Department of Business Administration, North Borneo University College, Kota Kinabalu, Sabah, Malaysia

© The Author(s), under exclusive license to Springer Nature Switzerland AG 2021   55
A. E. Hassanien et al. (eds.), *The Global Environmental Effects During and Beyond COVID-19*, Studies in Systems, Decision and Control 369,
https://doi.org/10.1007/978-3-030-72933-2_4

1 Introduction

The new type of coronavirus is a pathogen that causes severe and exceptional infectious pneumonia (COVID-19). In Geneva, the World Health Organization announced that China's pneumonia outbreak constituted a public health emergency on January 30, 2020. Taiwan's Ministry of Health and Welfare's Disease Control Agency announced on January 15, 2020, that severe particular infectious pneumonia had been added as the fifth category of notifiable infectious diseases. Most of the human coronavirus transmission is mainly transmitted by direct contact with viral secretions or droplets. Coronavirus infection in humans is primarily caused by respiratory symptoms, which include coughing, runny nose, nasal congestion, fever, and other general upper respiratory tract infection symptoms. However, a small number of more serious respiratory diseases, such as pneumonia, can cause serious death. For the response to the COVID-19, medical institutions, centralized quarantine sites, home quarantine, and the general public can adequately handle their waste. The operating principles of this waste classification management are specially formulated [1–3].

Initially, Taiwan's identification and classification of "infectious industrial waste" mainly referred to the "Medical Waste Tracking Act" of the United States in 1988. However, the infection risk of medical waste is not all the same, and the "sharp instrument waste" quickly causes invasive injuries and increases the risk of infection. Therefore, the World Health Organization (WHO) believes that sharp instrument waste must be specially managed regardless of whether they come into contact with infectious substances. The risk of "genotoxic waste" is not infectious. Its biochemical toxicity can cause mutations, deformities, or cancer. Internationally, hazardous chemical substances are controlled, and high-temperature incineration is required to destroy the chemical structure of substances and remove them' harmful properties. Therefore, when the Environmental Protection Agency of Taiwan revised the "Hazardous Industrial Waste Identification Standard" in 2006, it added "genotoxic waste" and distinguished "sharp instruments waste" from infectious waste, highlighting the difference from "Infectious waste" hazard characteristics to implement management. The definition of medical waste in countries and international organizations is defined as follows [4–6].

- The WHO definition of Healthcare waste: refers to waste generated from health care institutions, including waste generated from diagnosis, treatment, epidemic prevention, rehabilitation, and related research activities.
- The United Nations Environment Programme (UNEP) definition of biomedical and health care waste: refers to the solid and liquid waste generated by health care activities, including collected gaseous waste (waste canned gas).
- US Environmental Protection Agency (USEPA) medical waste: refers to all waste generated by medical facilities, including hospitals, clinics, physician offices, dentists, blood banks, veterinary hospitals/clinics, medical research facilities, and laboratories. Categories requiring special management include blood-stained bandages, microbial culture dishes, surgical gloves, surgical instruments,

discarded needles, culture strains, and inoculation loops, removed human organs, and lancets etc.
- The definition of clinical waste in the United Kingdom: waste includes or part of human or animal tissues, blood, body fluids, excrement, drugs, gauze, clothing, syringes, needles or other sharp instruments, etc., contact will endanger personal safety Waste.
- Taiwan's definition of biomedical waste: Refers to medical institutions, medical laboratories, medical laboratories, industrial and research institutions with bio-safety level 2 or higher laboratories, laboratories engaged in genetic or biotechnology research, biotechnology factories, and Pharmaceutical factories, waste generated during medical treatment, medical inspection, post-mortem, quarantine, research, pharmaceutical or biological material manufacturing, including genotoxic waste, waste sharp instruments, and infectious waste.

According to surveys and studies in various countries worldwide, biomedical waste (or infectious waste) accounts for about 10 to 15% of the total hospital waste taking Taiwan's "Waste Disposal Law" as an example. Waste is divided into industrial waste and general waste. The waste industrial includes general industrial waste and hazardous industrial waste. Hazardous industrial waste is defined as toxic and dangerous waste and whose concentration or quantity is sufficient to affect human health or pollute the environment. Besides, it is divided into Standards for Defining Hazardous Industrial Waste. Substances listed as regulated hazardous industrial waste are "Other waste officially announced by the central competent authority" and "Hazardous industrial waste determined by hazardous characteristics." Biomedical wastes that are more likely to be infectious are classified as regulated "hazardous industrial waste". General industrial waste refers to industrial waste generated by medical institutions other than hazardous industrial waste. According to Standards for Defining Hazardous Industrial Waste, wastes that are unique to medical institutions and require special management (i.e., biomedical wastes) can be subdivided into three categories (see Fig. 1 for details) [7, 8]:

- Genotoxic waste: cytotoxins or other drugs that cause cancer or may cause cancer.
- Waste sharp instruments: waste products that can cause stab or cut injuries to the human body.
- Infectious waste: 9 wastes, including waste microbial cultures.

The hospital's management of the waste generated is divided into two main parts, and this is shown below(Fig. 2):

(1) Infection control management standards (health agencies and hospitals): Refers to the internal management of the infection control system within the hospital, including controlling pollution that may arise from the operation of waste and is supervised by the health agency.

(2) Waste Disposal Law (Environmental Protection Agency): Refers to waste storage, removal, treatment, flow tracking management, special emphasis on

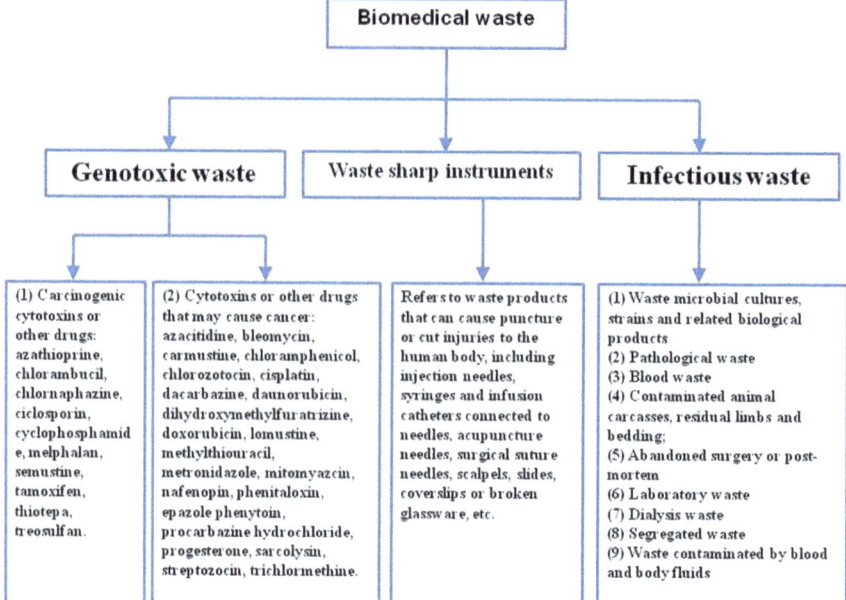

Fig. 1 Classification of biomedical waste

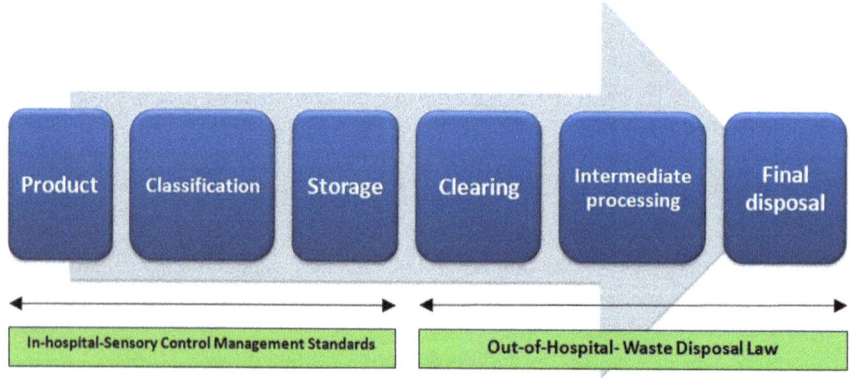

Fig. 2 Rights and responsibilities of medical waste management

the classification and centralized storage of waste in the hospital, and the tracking and proper disposal of waste flow leaving the hospital.

According to the management methods of waste cleaning professional technicians in Taiwan, professional technicians have trained and qualified to complete the application to set up in accompany. The categories of experienced technicians include: (1)

Waste removal technicians divided into Class A, Class B, and Class C; and (2) Waste Processing technicians are classified into Class A and Class B. The Environmental Protection Agency has announced "Designated Announcement of Businesses for Waste Professional and Technical Personnel," that designated medical institutions should set up the types and number of professional and technical personnel. Medical centers, regional hospitals, and hospitals with more than 50 patient beds should have persons 1 class B or above waste processing technicians. If they remove or dispose off hazardous industrial wastes by themselves, a Class A processing technician should be assigned [9, 10].

The storage method of sharp instrument waste and infectious waste of biological, medical waste, and sharp instrument waste should be stored separately from other wastes and sealed in a substantial container that is not easily penetrated, which has its storage limited to one year (Fig. 3). Infectious wastes should be stored separately from other wastes; those treated by heat should be sealed in leak-proof, non-breakable red plastic bags or red flammable containers. Those treated by Sterilization should be stored in leak-proof and non-breakable sealed yellow plastic bags or yellow containers (Fig. 4). Storage conditions should meet the following requirements.

(1) Waste-producing institutions: for storage at a temperature above 5 °C, the limit is one day; for refrigeration below 5 to 0 °C, the limit is seven days; for freezing below 0 °C, the limit is 30 day.

(2) Removal agency: no storage; however, if there are special circumstances that require transshipment, it can be refrigerated or frozen at less than 5 °C with the approval of the competent local authority, and the limit should be 7 days.

(3) Handling agency: not to store above 5 °C; for cold storage below 5 to 0°C, the limit is seven days; for freezing below 0 °C, the limit is 30 days.

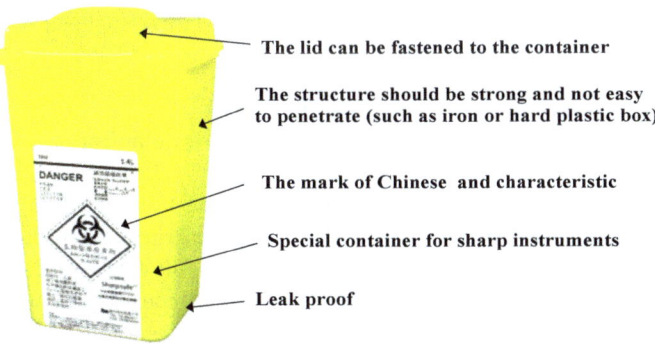

The lid can be fastened to the container

The structure should be strong and not easy to penetrate (such as iron or hard plastic box)

The mark of Chinese and characteristic

Special container for sharp instruments

Leak proof

Fig. 3 Schematic diagram of storage container for waste sharp instruments

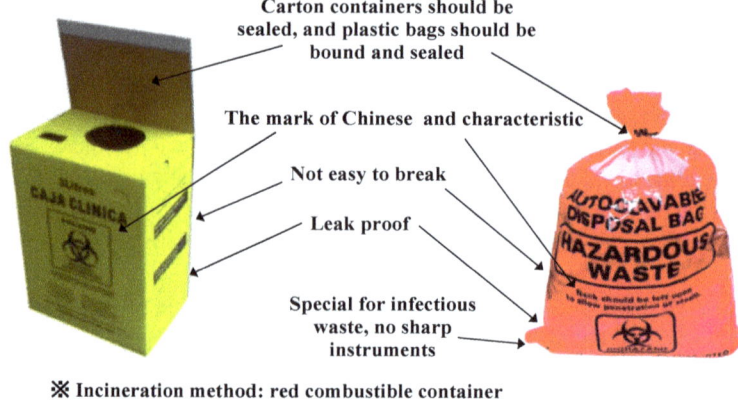

Carton containers should be
sealed, and plastic bags should be
bound and sealed

The mark of Chinese and characteristic

Not easy to break

Leak proof

Special for infectious
waste, no sharp
instruments

※ Incineration method: red combustible container
※ Sterilization treatment: yellow container

Fig. 4 Schematic diagram of the infectious waste storage container

2 Waste Classification and Cleaning Methods in COVID-19 Medical Institutions

Regarding the waste classification and cleaning methods of COVID-19 medical institutions and the recommendations of the "Guidelines on Control Measures for Medical Institutions Responding to COVID-19 Infection", the relevant regulations of the "Waste Cleaning Law" issued by the Environmental Protection Agency in Taiwan should still be followed for disposal [11, 12].

(1) The waste generated by medical institutions includes general waste, general business waste, and hazardous business waste (biomedical waste). The classification description is as follows:

   (1.1) *General waste is summarized in* Table 1.
   (1.2) **General business waste** such as medical equipment packaging materials, drip bottles (soft bags), medicine bottles, toilet paper, diapers, small medicine bottles, ampoules, and masks used by the public.
   (1.3) **Biomedical waste** such as sharp waste instruments, infectious waste, etc. The types are roughly described as follows. For detailed classification, please refer to the "Hazardous Industrial Waste Identification Standard-chapter of Biomedical Waste" set by the Environmental Protection Agency in Taiwan (Table 2).

   The cleaning process of waste generated in the hospital is classified, packaging, storage, removal, intermediate treatment, and final disposal. After various types of waste leave the hospital, the main cleaning flow is as follows:

(1) General industrial waste: it is mainly commissioned by the cleaning and disposal company to carry out the removal and treatment. The treatment

Table 1 The classification description summarized of general waste

Item	Classification description
General waste	Domestic waste is generated by visitors and employees, such as paper, plastic bags, toilet paper, disposable chopsticks, etc.
According to the "General Waste Recovery and Disposal Measures", general waste must be classified in the following ways before it can be recycled, removed or processed	1. Huge garbage: refers to large-scale discarded items such as furniture, pruned garden branches or general waste announced by the competent authority 2. Resource Garbage: refers to the general waste recycling items (except food waste) announced in accordance with Article 5, Paragraph 6 of the "Waste Disposal Law" and items that should be recycled in accordance with Article 15 Paragraph 2 of the "Waste Disposal Law" or General waste generated from its packaging and containers after being eaten or used 3. Hazardous waste: refers to general waste that meets the identification standards for hazardous industrial waste and has been announced by the central competent authority 4. Food waste: discarded raw and cooked food and its residues or organic waste announced by the competent authority 5. General waste: refers to general waste other than huge waste, resource waste, hazardous waste, and kitchen waste

Table 2 Biomedical waste type description collection

Item	Classification description
Waste sharp instruments	Waste products can cause stab or cut injuries to the human body, including injection needles, syringe barrels and infusion catheters connected to surgical suture needles, acupuncture needles, general needles and scalpels, glass slides, coverslips, or broken glassware, etc.
Infectious waste	Divided into waste microbial cultures, strains and related biological products, pathological wastes, blood wastes, contaminated animal carcasses, residual limbs and bedding, surgical or post-mortem wastes, laboratory wastes, and dialysis waste, segregated waste, waste contaminated by blood and body fluids. Personal protective equipment (PPE) produced from isolation wards or due to high-risk disposal processes such as intubation, diversion visits or admission wards, regardless of whether it is contaminated with patient blood or body fluids, shall be in accordance with "Hazardous Industrial Waste Identification Standard-chapter of Biomedical Waste "and other regulations, should be recognized as infectious waste

methods include incineration, landfill treatment, and reuse of industrial waste. The ash after incineration is treated by landfill.

(2) Bio-medical waste: among the hazardous industrial waste generated by hospitals, bio-medical waste is the main item. The general cleaning flow includes commissioning an agent for treatment, treatment by a commontreatment institution, or treatment by medical institutions, most of which are (97%) incineration and heat treatment were used. After incineration, the ash is buried or reused. Among them, "infectous waste" is treated by heat treatment in principle. Some items are destroyed after Sterilization and can be treated as general industrial waste for sharp instrument waste. They are treated by heat treatment or crushed after Sterilization (Fig. 5).

(3) Genotoxic waste: Treated by heat treatment or chemical treatment. The flow of "genotoxic waste" cleanup includes commissioned treatment companies, joint treatment institutions, or hospitals. The treatment methods are heat treatment or chemical treatment. After incineration, the ash is buried.

Following the Taiwan Environmental Protection Agency's "Industrial Waste Storage and Clearance Treatment Methods and Facilities Standards" and the Department of Health "Announcement of Certain Infectious Medical Waste Sterilization Standards and Related Regulations," the incineration treatment and sterilization treatment of hazardous and biomedical waste are sorted out as shown in the Table 3.

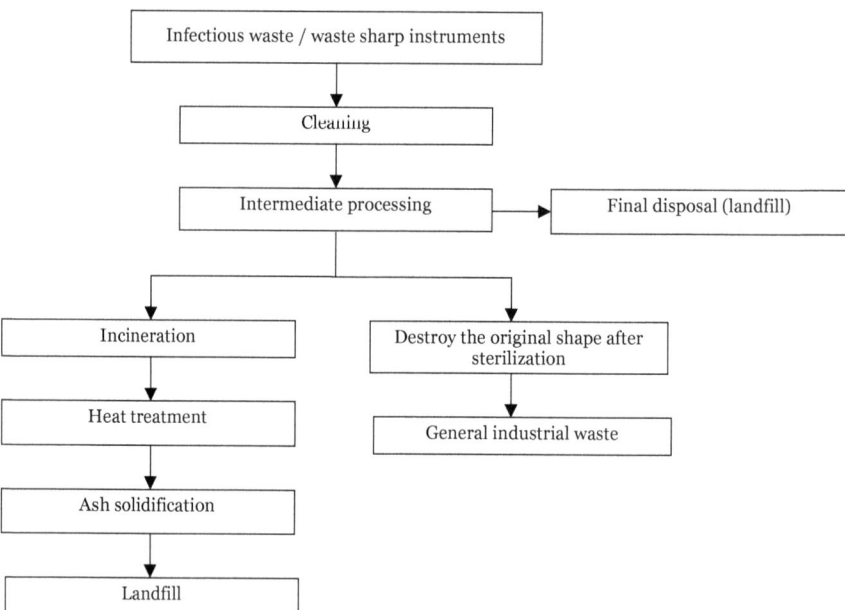

Fig. 5 The flow of biomedical waste cleanup

Table 3 Incineration and sterilization conditions for hazardous and biomedical waste

	Hazardous industrial waste	Biomedical waste
Incineration	The temperature of the exit center of the combustion chamber: above 1000 °C It is burning gas retention time: more than 2 s Destruction and removal efficiency of harmful organic substances such as organic chlorides: over 99.99%. Polychlorinated biphenyls (PCBs) and dioxin: more than 99.999% Toxic chemical substances: more than 99.9%	• The temperature of the exit center of the combustion chamber: above 1000 °C • It is burning gas retention time: more than 1 s • Combustion efficiency: over 99.9%
High temperature and high-pressure steam sterilization	(Not applicable)	• Operating conditions include the following: 1. The temperature needs to exceed 121 °C, the pressure reaches more than 1.06 kg/cm², and the heating time needs to exceed 60 min 2. The temperature is above 135 °C, the pressure is above 2.18 kg/cm² and the heating time is above 45 min • Sterilization efficiency test uses Bacillus subtilis spores or thermophilic bacillus spores as the test and the sterilization efficiency can be up to 99.999% • Mechanical testing includes continuous monitoring of temperature, pressure, and time • The chemical test is to use test paper or test tape or steam clock for the sterilized material for each operation (i.e. each pot) • The biological test is to use the thermophilic bacillus spore biological test bottles (biological culture seedlings) to be placed in the pot and sterilized container at least once a month to complete the biological test • The filling rate should not exceed 80% of the total volume of the autoclave • Destruction of the prototype after Sterilization refers to discarded microbial cultures, strains and related biological products, surgical or post-mortem wastes, laboratory wastes, dialysis wastes, wastes contaminated by blood and body fluids • Crushing after Sterilization is a waste sharp tool

3   COVID-19 Centralized Quarantine, Home Quarantine, and Isolation Waste Cleaning Methods

To avoid the spread of COVID-19 infectious disease, the Environmental Protection Agency in Taiwan will handle the waste generated by centralized quarantine and quarantine following high infectious waste standards. After the Central Epidemic Command Center in Taiwan provides the list, it will be assigned to a level A waste removal agency to deal with the list.

Centralized quarantine station waste cleaning methods include the following.

3.1   Waste Packaging

– The waste produced by the quarantined person should be packed in garbage bags, and the openings of the garbage bags should be sealed. The garbage must be properly packaged and placed at the house's door at the agreed collection time by the regulations of the centralized quarantine site.
– If there are sharp objects, they should be wrapped with papers like newspapers to prevent the sharp objects garbage bag from piercing the garbage bag's surface. If the bag leaks, it should be filled with a second garbage bag immediately.
– Avoid squeezing the garbage bags containing waste, and seal the garbage bags' openings when the garbage bags are 70% full.
– If foul odor occurs, the garbage bag should be sealed first, and then the second layer of the garbage bag should be used for loading.
– Garbage discharge: limit the waste generated in daily life and must not contain huge garbage.

3.2   Other Matters Needing Attention

– The waste is cleared and transported by vehicles of the Class A waste removal agency commissioned by the Environmental Protection Agency in Taiwan.
– It will be cleared every two days, and outlying islands will be cleared every seven days. The frequency of clearing and the method of collection and transportation should be adjusted to local conditions.
– The waste produced by the personnel stationed in each centralized quarantine site (such as police or staff) and the separated should be stored, collected, and transported separately. The waste produced by the stationed personnel should be disposed of by the original receiving and transporting unit of the site.
– It was summarized for waste collection procedures as shown in Table 4.

The waste disposal methods of people under home quarantine are described below. In accordance with the relevant provisions of the "Isolation Measures for Violation

Table 4 Biomedical waste type description collection

Oder	Waste collection procedures description
1.	The collection method should be adapted to local conditions. The personnel of the Class A waste removal agency in Taiwan shall collect them at the concentration points or collect them one by one at the door of the quarantined house
2.	When collecting, make sure that the garbage is sealed in the bag
3.	In principle, waste collectors should not enter the living space of quarantine persons and avoid close contact

of the Provisions of Article 48 Paragraph 1 of the Infectious Disease Control Act, the Standards of Penalty for Cases of Quarantine Measures in Article 58 Paragraph 1, 2 and 4", the person in home isolation/quarantine period are not allowed to go out and throw out the trash. The garbage produced in daily life, such as used masks, toilet paper, and other items should be packed in garbage bags with the bag's openings properly sealed, properly collected, and temporarily stored at home. Suppose there is a need for garbage disposal. In that case, the Environmental Protection Department can provide special telephone lines for each county and city. The family members will seal it up and send it to the cleaning and disposal company contracted with the Environmental Protection Agency in Taiwan or after the quarantine is over, the person who is freed from the COVID-19 can send the well-sealed garbage to the garbage truck of the cleaning team to be collected and transported in accordance with the usual garbage disposal methods. Those who live alone or have limited mobility and cannot dispose of rubbish during the home isolation/quarantine period can do that when the quarantine is over and freed. This is because the public may become infected with COVID-19, and the rubbish produced will be treated as garbage produced by the infected person of COVID-19 and shall be treated as infectious waste. The local environmental protection bureau must be contacted first, and the waste removal agency of Class B (or Class A) will clear and transport the garbage to the concentration point. The Class A waste removal agency commissioned by the Environmental Protection Agency in Taiwan is transferred to the Class A waste treatment facility to avoid the spread of infectious diseases.During home quarantine or home isolation, the garbage packaging proceduresrefer to Table 2 Centralized quarantine or home isolation/quarantine waste disposal methods.

Matters needing attention for garbage removal staff

- If there are special circumstances, garbage must be discarded during the stay at home. The garbage removal staff shall not enter the isolated/household living space in principle and avoid close contact.
- Garbage removal staff are recommended to wear medical masks and gloves to perform business operations and also make sure that the garbage bags are sealed when collecting garbage.
- For information about the garbage collection, transportation, and disposal operations for people in-home quarantine, please refer to the Special Severe Special Infectious Pneumonia Special Project of the Environmental Protection Agency in

Taiwan "Environmental protection agencies at all levels handle garbage collection, transportation, and disposal operations for people who are in-home quarantine for a program of response to severe special infectious pneumonia" (Table 5).

In addition to complying with the requirements above, the removal methods of biomedical wastes are also by Article 18 of the "Standards for Storage, Removal, and Treatment of Industrial Wastes and Facilities," and the requirements are as follows (Figs. 6, 7) [13, 14]:

1. Waste stored in containers of various colors shall not be mixed and removed.
2. During transportation, it cannot be compressed and opened arbitrarily.
3. Refrigeration measures should be provided during transportation, and normal operation should be maintained.
4. If no staff is present during the loading and unloading process, the cleared vehicle's door should be kept closed and locked.

4 COVID-19 Prevention of Epidemic Infection Control and Risk Assessment Measures in Taiwan Hospitals

4.1 Epidemic Prevention and Infection Control in Hospitals in Taiwan

– In response to the COVID-19 outbreak and current pneumonia and influenza epidemic season and to prevent the spread of the epidemic in medical institutions, the Taiwan Disease Control Bureau stated that it is necessary to be vigilant in medical practice andconsider the following measures (Table 6).

Medical institution personnel takes care of cases of special infectious pneumonia. At this stage, it is recommended to follow the principles of standard protective measures, droplet infection, contact infection, and airborne infection protection measures. Also, to adopt appropriate infection control and protection measures such as (Fig. 8):

In response to the severe and special infectious pneumonia epidemic, supervised district hospitals must strengthen the implementation of "TOCC" consultations and patient diversion to see infection control measures (shown in Table 7).

Epidemic prevention risk control and health management measures

Establish a hazard risk control mechanism to ensure the safety and health of employees. During the epidemic period, in addition to cooperating with the epidemic prevention measures of the Department of Disease Control of the Ministry of Health and Welfare in Taiwan, the hospital should take necessary risk control and health management measures in response to the development of the epidemic and the needs of employee safety protection. This includes the following:

Table 5 Waste disposal methods for centralized quarantine or home quarantine

	Huge garbage	General garbage	Kitchen waste
Type of waste	Large-volume waste furniture or general waste announced by the competent authority	Masks, toilet paper, disposable chopsticks, dirty plastic bags, lunch boxes, and other daily garbage	Discarded raw and cooked food and its residues or organic waste
Processing method	During the home isolation/quarantine period, non-daily garbage containing huge garbage and home appliances will not be collected	1. Garbage does not need special classification, it should be packed in garbage bags, and the mouth of the bag should be sealed to avoid exposure of garbage 2. If there are sharp objects, they should be wrapped in newspaper to prevent the garbage bag from piercing the surface. If the bag leaks, it should be filled with a second garbage bag immediately 3. Avoid squeezing the garbage bags that contain garbage, and seal the mouth of the garbage bag when the garbage bag is 70% full 4. For odor-prone garbage, it is recommended to seal the garbage bag first, and then add a layer of garbage bag to seal it tightly, and separate it with other garbage 5. During the home isolation/quarantine period, people are not allowed to go out and throw out garbage by themselves 6. If there is a need to discard them during the home isolation/quarantine period, because people may become infected with COVID-19, the standard treatment of infectious waste must be contacted with the local environmental protection bureau	1. Drain the food waste and pack it properly, and put it in a specific floor of a personal refrigerator. It is ideal if there is a frozen layer 2. If the food waste in the bag leaks or odor occurs, it should be filled with the second layer of garbage bag immediately

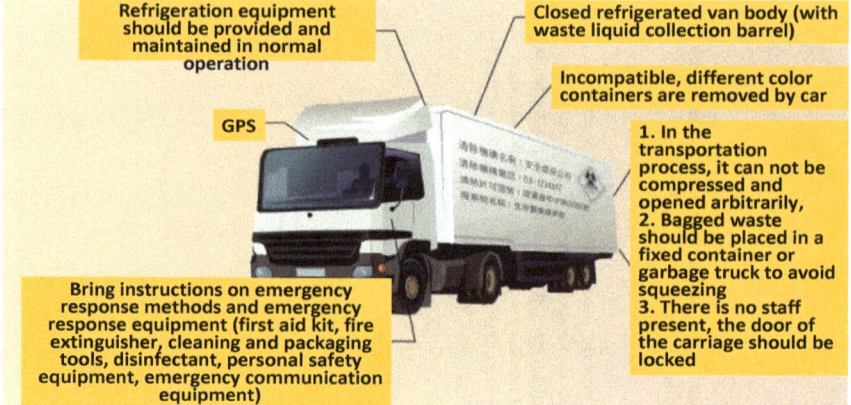

Fig. 6 Schematic diagram of waste sharp instruments and infectious waste removal vehicle specifications

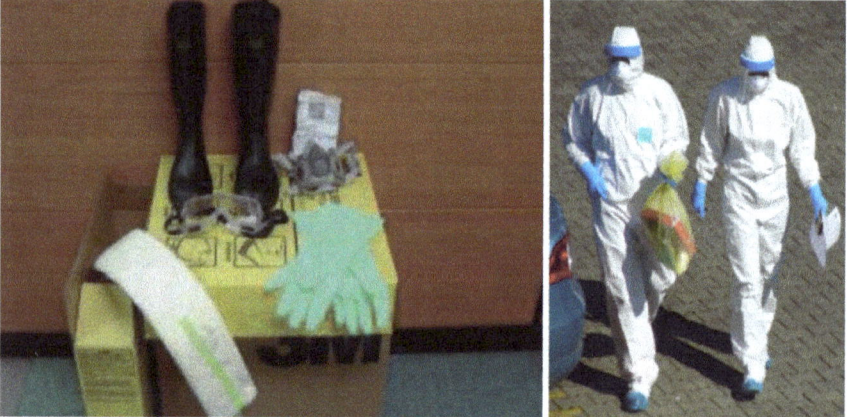

Fig. 7 Schematic diagram of PPE carried with the vehicle

- Proper and sufficient masks should be provided, and workers should not be prohibited from wearing them.
- Establish body temperature measurement and screening measures to strengthen employee health management.
- Strengthen the cleanliness and ventilation of all areas of the workplace. The floors, walls, appliances, and articles should be disinfected in places where the infection is likely.
- Strengthen workplace infection prevention and labor, self-protection education, and training.

Table 6 Summary of medical vigilance and measures to be taken in Taiwan

Oder	Medical vigilance and measures description
1.	The following "TOCC" mechanism should be implemented for outpatients in Taiwan hospitals with emergency checkups and suspected respiratory tract infections. They should inquire and record travel history, occupation, contact history and whether they are clustered or not. Also, relevant infection control measures are followed for appropriate isolation protection
2.	Offering or assisting patients with fever or respiratory symptoms to wear surgical masks in outpatient and emergency areas
3.	The outpatient and emergency areas should have a mechanism for diversion of patients. The clinics or examination rooms in the outpatient and emergency departments as the triage area should be well ventilated. The waiting room should be well ventilated and the patient's flow should be properly arranged and the use of assessment, diagnosis and treatment or inspection should be provided
4.	Post obvious notices in outpatient and emergency areas and also at the entrance of the hospital to remind people seeking medical care to wear their mask. Patients with fever or respiratory symptoms must be made to wear a mask whiles waiting for consultation. Relevant information must be given to the doctor whether or not these patients have visited a high-risk area of epidemic recently. Information such as regional travel history of the patient must be provided so that doctors can establish correct diagnosis
5.	Medical care workers should strictly follow respiratory hygiene and cough etiquette. If symptoms such as coughing and sneezing are realized among them, the wearing of masks and hand hygiene throughout the course of caring for patients must be strengthened. If a worker is having fever and acute respiratory symptoms, the worker should be recuperated at home and take the initiative to report to the unit supervisor to implement health monitoring and management
6.	If the patient diagnosed and treated with severe pneumonia has traveled to a high-risk epidemic area within 10 days before the onset of illness, they can go to the infectious disease notification system of the Department of Disease Control and select "Other" and fill in the "Other Disease Name" column. Unexplained pneumonia history of travel in high-risk areas," and throat wipes, sputum and serum are collected and sent to the Disease Control Agency in Taiwan for examination

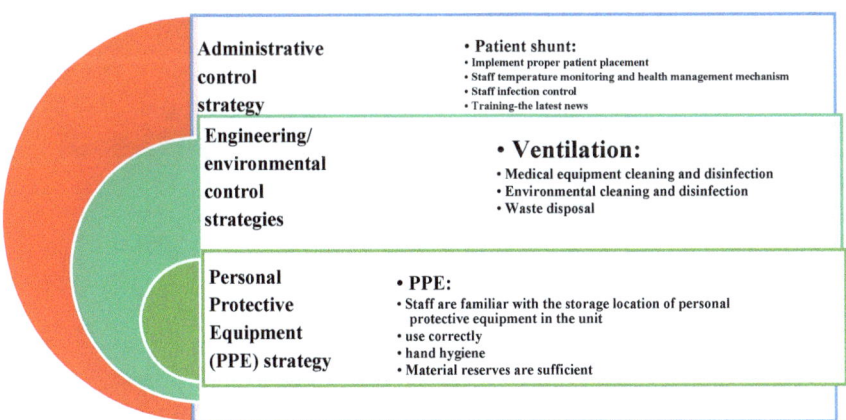

Fig. 8 Severe special infectious pneumonia-nosocomial infection control measures

Table 7 Summary of medical vigilance and measures to be taken in Taiwan

Item	Medical vigilance and measures description
Administrative control strategies-case notification and placement	**Hospital clinic** • Patients who meet the notification should be placed in an independent consulting room for evaluation. • Before entering the clinic, staff should wear appropriate PPE.
	Hospitalized • Priority is given to the single-person negative pressure isolation ward. While waiting for the negative pressure isolation ward, it may be temporarily placed in a single-person ward with sanitary equipment. • If there is no negative pressure isolation ward, the patient should be transferred to a hospital with a negative pressure isolation ward as soon as possible. • Do not use a positive pressure isolation ward, and the door of the ward should be kept closed. • Patients are treated according to the doctor's clinical disease treatment, and if necessary, the commander of the infectious disease prevention and treatment network will be under the jurisdiction. • Taking into account the privacy of the patient, indicate the isolation and protection measures that the patient needs to take at the door of the ward, and only allow necessary personnel to enter the isolation room.
	Referral • If a suspected case requires a referral, a clear transfer procedure should be carried out with the receiving hospital first. • If you need to be referred to the emergency hospital in the network area, the commander of the infectious disease prevention and treatment network must be instructed. • If there are accompanying medical care personnel, they should wear appropriate PPE.
Administrative control strategies-(visiting) visitor management	1. The number of visitors should be limited 2. If visitors still have necessary reasons to enter the ward, they should be taught how to properly use PPE and perform hand hygiene 3. Visitors can enter the ward after wearing various PPE as recommended 4. Keep all visitor records, including name, contact number and address

(continued)

Table 7 (continued)

Item	Medical vigilance and measures description
All staff and visitors entering the isolation ward should wear the following PPE	1. Long-sleeved waterproof isolation clothing: It is recommended to use disposable long-sleeved waterproof isolation clothing
	2. Surgical gloves
	3. High-efficiency filter masks of N95 or equivalent (inclusive): The tightness inspection of the mask must be performed before each use
	4. The place where PPE is worn should be equipped with a close inspection
	5. When performing medical treatments that may cause droplet particles: wear high-efficiency filter masks (N95 or equivalent), gloves, gowns, goggles or face shields, and hair caps as needed
	6. Perform in a negative pressure room or a well-ventilated single room
	7. Only necessary personnel are allowed to stay in the ward to reduce the number of exposed persons
	8. Related medical treatments such as: end tracheal insertion (extraction) of sputum, bronchoscope, induced sputum treatment, use of face-mask positive pressure respirators, nasal irrigation, nasal/throat/nasopharyngeal swabs and other removal of individuals after PPE and hand hygiene must be perform immediately
	(1) PPE wear and removal process and hand hygiene should be performed after each step in the removal process
	(2) The removed PPE should be thrown into the medical waste trash can. If it is reusable, it should be placed in a designated container for subsequent disinfection
	9. In response to the severe pneumonia epidemic, PPEs are recommended for medical care workers

(continued)

Table 7 (continued)

Item	Medical vigilance and measures description
Hand hygiene	1. Always follow the 5 precautions of hand hygiene such as before and after contact with the patient, after removing PPE and after cleaning the environment 2. Use soap and water or alcohol-based dry-cleaning hand sanitizer for hand hygiene as appropriate 3. Staff should not wear rings, watches, and any wrist accessories

(continued)

Table 7 (continued)

Item	Medical vigilance and measures description
Environmental/engineering control strategy–environmental cleaning and disinfection	1. Environmental cleaning and disinfection personnel: should receive appropriate training and wear appropriate PPE when performing work as recommended
	2. The other areas of the disinfection ward should be cleaned first, and then the isolation ward should be cleaned and disinfected
	3. Environmental cleaning and disinfection must start in the low-polluted area and end in the heavily-polluted area
	4. After stopping the medical treatment that produces droplets, it takes about 12–15 air changes per hour for 20 min before entering the space to perform environmental cleaning and disinfection
	5. Cleaning utensils should be cleaned and disinfected after each use and replaced in due course according to the condition after use
	6. When performing cleaning work, first use detergent or soap and water to remove dirt and organic matter, and then use a damp cloth and a suitable disinfectant to perform effective environmental cleaning and disinfection
	7. The disinfectant should be used in accordance with the dilution method, contact time and treatment method recommended by the manufacturer or the bleach dilution made on the same day. Surfaces frequently touched by patients (such as bedside tables, bedside tables, bed rails and other ward furniture) should be cleaned daily and disinfected with appropriate disinfectants or 1:50 diluted bleach (1000 ppm). The surface of the bathroom or toilet should be cleaned daily and disinfected with 1:10 diluted bleach (5000 ppm) as follows
Environmental/engineering control strategy–medical waste	1. All waste generated in the isolation ward/area should be discarded in appropriate containers or bags to ensure that it will not overflow or leak
	2. Personnel handling waste should wear appropriate PPE as recommended
	3. Waste should be processed in accordance with the relevant regulations of the "Waste Removal Law" issued by the Environmental Protection Department of the Executive Yuan in Taiwan
Infection control measures–sample handling	1. All specimens must be considered as biohazard
	2. When transferring, it should be packed in double-layer zipper bags and marked with biohazard labels
	3. Use manual delivery of the specimen; do not use pneumatic-tube systems for delivery
	4. Laboratory operation level: Biosafety level 2 or higher: clinical specimen processing and PCR inspection operations, etc.
	5. Biosafety level 3 or higher: pathogen isolation, identification, cultivation or high-concentration, large-scale operation of pathogens, etc.
	6. Steps that may produce infectious aerosol should be carried out in a biological safety cabinet (BSC)
	7. For other related suggestions, please refer to the "Laboratory Biosafety Guidelines for the Handling of Severe Special Infectious Pneumonia Samples in Medical Laboratories"

(continued)

Table 7 (continued)

Item	Medical vigilance and measures description
Infection control measures–critically ill patients	1. Respirator: It must have a high-efficiency filter device and be cleaned and disinfected according to standard procedures after use
	2. Try to use disposable breathing apparatus tubing. If other reusable medical devices must be used, they must be disinfected according to the product instructions
	3. Avoid using non-invasive positive pressure breathing apparatus
	4. A closed suction system should be used where necessary, the integrity of the respirator circuit should not be compromised
	5. When performing respiratory care, cough-inducing activities or drug spray treatment, only necessary and properly protected medical personnel are allowed to enter

- The health and safety of workers is the top priority. If not necessary, workers should avoid going to China, Asia Pacific, Europe, the United States, and other declared epidemic areas.
- For the management of high-risk employees, emergency response strategies and procedures should be established for high-risk units such as medical laboratories handling severe and special infectious pneumonia specimens.
- For front-line workers who will come into contact with an unspecified number of people and are at risk of infection, first-line workers shall implement necessary protective measures and provide appropriate PPE (such as medical masks) and take occupational safety and health measures. The disinfection of the work environment and cautioning workers to take personal epidemic prevention measures should be strengthened.
- Employers should strengthen cleaning, disinfection, and ventilation for confirmed cases of recently engaged in work or workplaces. Simultaneously, for workers who have been in contact with or are at risk of infection, appropriate PPE (such as N95 masks) must be provided and must be ensured to be used. It is determined that suspected or confirmed cases should be placed in special wards, equipped with a negative pressure isolation environment, and disinfection operations shall be implemented in accordance with the regulations of the central health authority (the infection control manual shall be disinfected according to the statutory infectious disease disinfection standards.)
- For workers who are likely to be exposed to biological pathogens' hazards, appropriate PPE should be provided for them and must be used. Articles contaminated by biological pathogens should be disinfected, stored and appropriately marked; utensils and supplies used in the related premises must be cleaned and disinfected.

5 Feasible Application of Smart System in Health Waste Management

Based on the complete COVID-19 waste management laws and principles, this article suggests that Smart System in Health Waste Management can be implemented according to Fig. 9. The summary description of each application scheme is shown in Table 8 [15, 16].

6 Conclusions

The hospital's epidemic prevention and infection control and risk management response measures are based on the "Medical Institutions' Response to COVID-19 Infection Control Measures" announced by the Central Epidemic Command Center in Taiwan. Risk-exposure operations should be planned as soon as possible to prevent disaster and mitigation and prepare measures to ensure safety and health

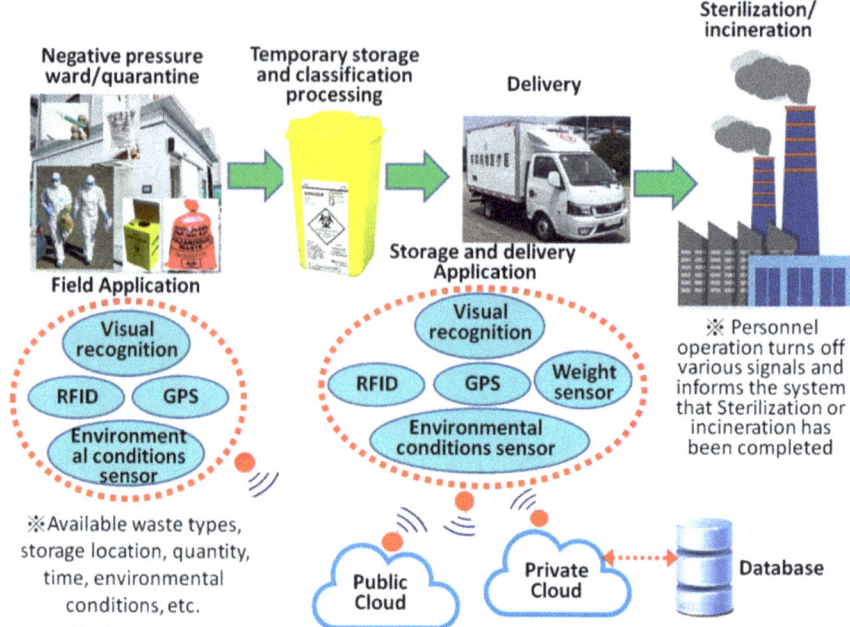

Fig. 9 The suggested architecture of smart system in health waste management

at work. Commencing the epidemic prevention and infection control plan of medical institutions, the following must be considered:

– Outpatient and emergency areas should have a mechanism for patient diversion to see a doctor.
– Infectious disease case notification system and handling.
– Principles of List of Contacts in Medical Institutions.
– Health monitoring and management of medical care workers
– Accompanying visit management.
– Personal Protective Equipment (PPE).
– Environmental cleaning and disinfection.

Following the principles and recommendations of standard epidemic prevention and infection control measures, droplet infection, contact infection, and air infection prevention measures, the response and prevention risk management and control measures provide hospitals concerning ensure timely response when the situation occurs and protect the wellbeing of patients and medical staff health. In response to the need for emergency surgery during the hospitalization of COVID-19 patients, it is necessary to establish appropriate hospitalization and surgical infection control measures for COVID-19 patients. Implementing occupational safety and health related laws and regulations and strengthening the ability to prevent the exposure of biological pathogens, effectively prevent the spread of the epidemic in hospitals

Table 8 Summary description of each smart system application scheme

Item	Smart system	Possible applications
Negative pressure ward/quarantine field application	1. Visual recognition 2. RFID 3. GPS 4. Environmental conditions sensor	1. Visual recognition can carry out personnel control, and carry out waste input identification, direct identification and control in classification, packaging bags/boxes, etc. 2. RFID can be installed in the packaging bag/box, so that the system can directly confirm the waste status, and can compare the time and other data in the database 3. GPS allows the bag/box to know its specific location and record its movement trajectory. If cloud computing can be performed, it can be determined whether the trajectory meets expectations 4. Environmental conditions sensor includes temperature, humidity and other sensors, because temperature and humidity affect the storage time of waste, it plays an important role in the actual management and control of waste storage and processing
Temporary storage and classification processing	1. Visual recognition 2. RFID 3. GPS 4. Environmental conditions sensor	1. Visual recognition can control in classification, packaging bags/boxes, etc. 2. RFID can directly confirm the waste status 3. GPS allows the bag/box to know its specific location and record its movement trajectory 4. Environmental conditions sensor includes temperature, humidity and other sensors, can be set in closed storage
Delivery application	1. Visual recognition 2. RFID 3. GPS 4. Environmental conditions sensor 5. Weight sensor	1. Visual recognition can identify the correctness of loading and unloading and the safety during transportation 2. RFID can obtain information about waste and transportation vehicles, and carry out loading and matching, and compare the correctness 3. GPS can obtain the transportation route and calculate the correctness of the route 4. The environmental conditions sensor mainly monitors the temperature and humidity of the conveying vehicle, which can ensure the safety of the conveying process and the variability of waste 5. The weight sensor can obtain the conveying weight, in addition to ensuring driving safety, it can also effectively control the amount of waste to avoid conveying errors
Sterilization/incineration application	Component recycling and shutdown	Personnel operation turns off various signals and informs the system that Sterilization or incineration has been completed

and workplaces. All necessary safety and health measures must be implemented to prevent medical staff and contract workers from developing infectious diseases of occupational disasters.The hospital's internal waste management must not only comply with the environmental protection laws and regulations but also take into account infection control and economic costs. Reference has been made to relevant international materials:

– The Technical Guidelines on the Environmentally Sound Management of Biomedical and Healthcare Wastes of the United Nations Environment Program(UNEP).
– The World Health Organization (WHO) "Safe Management of Wastes from Health-Care Activities" and collects domestic medical institutions' actual implementation experience. It collected relevant practical information and suggestions for waste management in the hospital to assist medical institutions in establishing a complete waste management system.

Finally, the smart systems such as Visual recognition, RFID, GPS, Environmental conditions sensor, Weight sensor can be applied in various stages such as negative pressure ward/quarantine field application, temporary storage and classification processing, delivery application, sterilization/incineration application. This application is not yet popular, and this gives it more room for improvement. This is also an essential prospect for smart systems in COVID-19 prevention, and it is worthy of continuous investment in development by all human beings.

References

1. Li, Q., Feng, W., Quan, Y.H.: Trend and forecasting of the COVID-19 outbreak in China. J. Infect. **80**(4), 469–496 (2020)
2. Chuand, K.C., Chang, K.C.: Hospital novel coronavirus (COVID-19) epidemic prevention and infection control and risk assessment measures. Ind. Safety Health **371**, 9–28 (2020)
3. Yang, Z., Zeng, Z., Wang, K., Wong, S.S., Liang, W., Zanin, M., Liang, J.: Modified SEIR and AI prediction of the epidemics trend of COVID-19 in China under public health interventions. J. Thor. Dis. **12**(3), 165 (2020)
4. Coronavirus disease (COVID-19) advice for the public. Geneva: World Health Organization. https://www.who.int/emergencies/diseases/novel-coronavirus-2019/advice-for-public (2020). Accessed 3 March 2020
5. Home care for patients with suspected novel coronavirus (COVID-19) infection presenting with mild symptoms, and management of their contacts: interim guidance, February 4 2020. https://www.who.int/publications-detail/home-care-for-patients-with-suspected-novel-corona virus-(ncov)-infection-presenting-with-mild-symptoms-and-management-of-contacts (2020). Accessed 3 March 2020
6. World Health Organization and the United Nations Children's Fund. 2020. "Water, Sanitation, Hygiene, and Waste Management for the COVID-19 Virus." Interim Guidance, April 23. WHO/2019-nCoV/IPC_WASH/2020.3. https://www.who.int/publications-detail/water-sanita tion-hygiene-and-waste-management-for-covid-19 (2020)

7. NESREA (National Environmental Standards And Regulations Enforcement Agency). 2020. "Guidelines for Handling Infectious Waste within the Context of Coronavirus (COVID-19)". April 16. Accessed May 26, 2020. https://www.nesrea.gov.ng/wpcontent/uploads/2020/02/Handling_Infectious_Waste_Guidelines.pdf (2020)

8. Rahman, M.M., Bodrud-Doza, M., Griffiths, M.D., Mamun, M.A.: Biomedical waste amid COVID-19: perspectives from Bangladesh. Lancet Global Health (2020)

9. Vanapalli, K.R., Sharma, H.B., Ranjan, V.P., Samal, B., Bhattacharya, J., Dubey, B.K., Goel, S.: Challenges and strategies for effective plastic waste management during and post COVID-19 pandemic. Sci. Total Environ. **750**, (2020)

10. Wang, J., Shen, J., Ye, D., Yan, X., Zhang, Y., Yang, W., Pan, L.: Disinfection technology of hospital wastes and wastewater: Suggestions for disinfection strategy during coronavirus Disease 2019 (COVID-19) pandemic in China. Environ. Pollut. 114665 (2020)

11. Cheng, Y.W., Sung, F.C., Yang, Y., Lo, Y.H., Chung, Y.T., Li, K.C.: Medical waste production at hospitals and associated factors. Waste Manag **29**(1), 440–444 (2009)

12. Yang, L., Yu, X., Wu, X., Wang, J., Yan, X., Jiang, S., Chen, Z.: Emergency response to the explosive growth of health care wastes during COVID-19 pandemic in Wuhan, China. Res. Conserv. Recycling **164**, (2020)

13. Mihai, F.C.: Assessment of COVID-19 waste flows during the emergency state in romania and related public health and environmental concerns. Int. J. Environ. Res. Public Health **17**(15), 5439 (2020)

14. Sangkham, S.: Face mask and medical waste disposal during the novel COVID-19 pandemic in Asia. Case Stud. Chem. Environ. Eng. **2**, (2020)

15. Kumar, A., Sharma, K., Singh, H., Naugriya, S.G., Gill, S.S., Buyya, R.: A drone-based networked system and methods for combating coronavirus disease (COVID-19) pandemic. Future Generat. Comput. Syst. **115**, 1–19 (2020)

16. Chang, K.-C., Chu, K.-C., Wang, H.-C., Lin, Y.-C., Pan, J.-S.: Agent-based middleware framework using distributed CPS for improving resource utilization in smart city. Future Generation Computer Systems **108**, 445–453 (2020)

Artificial Intelligence for Sustainable Waste Management and Control During and Post COVID-19 Crisis: Critical Challenges

Walid Hamdy⬤, Ashraf Darwish, and Aboul Ella Hassanien

Abstract COVID-19 swept the world. The total number of infected people has risen from 5 M in March 2020 to over 22 M in August 2020 and is increasing, which at the current stage does not seem to reach its peak. It has led to waste generation and various phases of waste management practice issues. The impacts include changes in the quantity of waste, composition, timing/frequency (temporal), distribution (spatial), and risk, which affect handling practices and care. So, the global dynamics of waste generation have changed, and special attention has therefore been required. The unexpected variations in the composition and amount of waste often need policymakers to react dynamically. This study underlines the challenges faced by the solid waste management sector during the pandemic and the prospects underlying the framework to resolve current lacunae. The study describes particular pharmaceutical waste cases, plastic waste, and food waste management, which were all a cause for great concern during this crisis. The combination of virus-packed biomedical waste with normal solid waste sources presents major negative protection and health challenges for sanitation employees without active citizen involvement and collaboration.

Keywords Waste management · Biomedical waste · Plastic waste · Food supply chain · Food waste · Solid waste management · COVID-19 waste

W. Hamdy (✉)
Faculty of Science, Port Said University, Port Said, Egypt
e-mail: walid.hamdy@ckes.cu.edu.eg
URL: http://www.egyptscience.net

A. Darwish
Faculty of Science, Helwan University, Cairo, Egypt
e-mail: ashraf.darwish.eg@ieee.org

A. E. Hassanien
Faculty of Computers and Artificial Intelligence, Cairo University, Giza, Egypt
e-mail: aboitcairo@cu.edu.eg

W. Hamdy · A. Darwish · A. E. Hassanien
Scientific Research Group in Egypt (SRGE), Cairo University, Cairo, Egypt

© The Author(s), under exclusive license to Springer Nature Switzerland AG 2021
A. E. Hassanien et al. (eds.), *The Global Environmental Effects During and Beyond COVID-19*, Studies in Systems, Decision and Control 369,
https://doi.org/10.1007/978-3-030-72933-2_5

1 Introduction

The new COVID-19 pandemic was rapid improvement [1], as can be seen from the statistics provided [2]. The sudden rise in requests for single-use goods and their use to protect the public population, service workers, health workers, and patients is one of the direct environmental impacts of that epidemic. As the epidemic causes major disruptions in the primary supply chain and concerns about waste management, the widespread use of protective equipment worldwide. The demand pattern is predictable to follow the global inclusive curve for different plastic items, such as personal protection equipment (PPE) and gloves and masks for health care professionals, plastic packaging parts for life-supporting devices, and rebreathers and general plastics materials, like syringes. Plastic materials used are also infected with pathogens and should be treated as harmful waste. Even previous the beginning of the COVID-19 pandemical, plastic waste management was believed to be a main environmental problem due to increasing concern about defilement in earthly and nautical environments [3]. The potential increase in waste volume from the COVID-19 pandemical is threatening to overwhelm the current waste management and the healthcare system. Global waste management schemes have also been incapable of coping satisfactorily with the current waste of plastic.

Medical waste from hospitals is especially troublesome because of removing any leftover pathogens [4]. Treatment plants are usually built to cope with steady-state environments under which medical waste is treated at a constant overall flow rate and composition. The thermal operation, including incineration and steam (auto-claving), the diverse treatment technology options are focused on plasma treatment and microwave treatment. The probability of treatment is determined by multiple economic, technical [5], environmental, and social acceptability. Rapid waste volume scale-up is likely to disturb processes designed for steady-state conditions [6]. A related issue is the decision to build new facilities to accommodate the increased amount of waste. Economics, pollution, protection, regulatory issues, and public acceptance are among the related aspects. It is, however, too late for those reflections at the start of the pandemic.

2 Fundamentals

During the emergency, as shown in Fig. 1, the ICU patients created about 6% of total medical waste, and the warning state states with no such waste stream before the emergency state. The possible infectious waste flow levels were more than 10 times lower than all emergency and warning states prior to the emergency state, leading to the rapid growth of COVID-19 incidents.

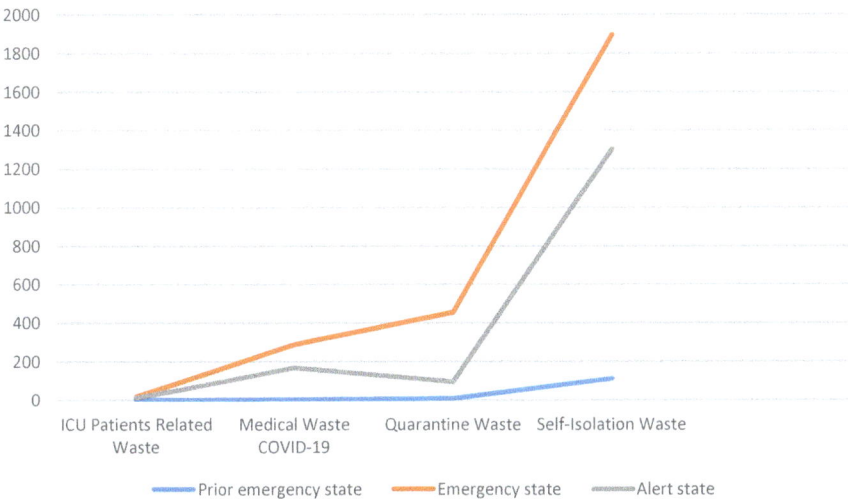

Fig. 1 COVID-19 waste flow

2.1 *Medical Waste Management*

Sustainable medical waste management, in particular, in the circumstances such as the COVID-19 pandemic, is problematic and exacerbated. Because of the global pandemic's novelty, adequate statistics on the quantity of manufactured medical waste, Waste generation hotspots and available treatment facilities are needed to modify existing waste facilities to manage unusual medical waste and the impact of the associated virus spread. Several technological technical knowledge in sorting, separation, transportation, storage, and probable waste management techniques are required to maximizing the existing infrastructure for a deal with emergencies [7] due to the possible rapid expansion of medical waste volumes. Improper medical waste management can expose accidents, illnesses, harmful effects, and air emissions to patients, healthcare staff and waste managers [8]. Nonhazardous waste, fulfilling waste, nuclear waste, contagious waste, chemicals waste, cytotoxic waste, sharp waste, and pharmaceuticals waste are the various sources of and derivatives of medical waste. A remarkable amount of recorded medical waste has contributed to the global pandemic, for instance, due to an increase in individual protective supply and instantaneous disposal after use. The COVID-19 pandemical in China has reportedly resulted in increased medical waste from personal protective equipment such as gloves, face masks, and eye protection [9]. Because of the whelming rise in everyday waste (i.e., over 240 metric tonnes) and the six-fold rise in hospital medical waste levels, the inflow of COVID-19 patients is reported to have led to the development of waste facilities and the designation in China of 46 mobile waste remediation plants [10]. Medical waste like masks, masks of total face, and gloves increased by 350%

in Barcelona, producing approximately 1,200 tonnes of medical waste instead of the normal waste of ~275 tonnes [11].

2.2 Nonmedical Waste Management

Increased development and use of nonmedical and household items such as a glove, thermometers, masks, cleaning products and toilet papers, foodstuffs, and sanitizers have been observed by the lockout institution stay-at-home politicking, and other preventative steps to control the spreads of COVID-19. Unexpected lockdown and scare of the virus have contributed to increased goods that single-use and scare purchases [12]. It is estimated that the matchless use of masks to decrease COVID-19 exposure has increased its production, thereby raising global revenues by US$ 166 billion. The COVID-19 pandemical appeared to had derailed attempts to decrease plastic waste due to the current function of protective devices such as single used gloves and masks. The WHO estimates a global monthly budget of 1.6 millones of plastic safety goggles, 76 millones of plastic test masks, and 89 millones of plastic surgical masks [13] to cover the distribution of COVID-19. In February, the normal plastic mask production in China increased by 116 million (ten times more than January) [14]. There are many estimates of tremendous plastic waste rising in Thailand from 1500 to 6300 tonnes per day due to food products distributed to homes. The UK saw a 300% increase in the illegal disposal of waste through the time of lockdown [15]. The pandemic of COVID-19 highlights the necessity for households to use a different set of garbage. The strict lockdown in Milan (Italy) resulted in a 27.5% decrease in overall waste production, including a 24.4% decrease in residual waste, a 20% decrease in paper and cardboard waste, a 16.7% lowering in waste glass, a 16.3% decrease in waste metal and waste plastic, a 14.4% decrease in domestic food waste and an 80.5% decrease in industrial food waste. Compared to other cities, the decline can be attributed to decreased waste generation. Waste recycling, however, increased by 1% relative to the same time in 2019, while street bins decreased by 38.2%. There was a 16.65% decrease in urban waste during the month of incarceration, from 282.3 thousand tons to 242 thousand tons in Spain (Catalonia). Due to mobile restrictions for tourism and business activities, waste generation declined by 25% in Barcelona. Print, lightweight cardboard, glass, lightweight packaging, and organic decreased by 20%, while mixed waste decreased by 12% [11].

3   Artificial Intelligence to Understand Waste Management

AI technologies transaction with the creation of computational systems and applications that can imitate human beings' attributes such as perception, learning, problem resolution, thinking, awareness of the environment, and understanding. The ability to resolve malady-defined problems, configure congregation mapping, and predict the performance, AI models like fuzzy logic (FL), artificial neural network (ANN), expert method, and genetic algorithm (GA) [16]. A specific role serves all AI models or sections of AI; for example, ANN models can train classification and prediction data. Besides, in urban geography, ANN's can be used to manage big-data and conduct spatial remote sensing. Besides having a knowledge base, human cognitive and analytical abilities can be developed by experience systems, like FL. Such systems have a basic linguistic syntax capable of handling complex activities and qualitative characteristics [16]. Evolutional like GA algorithms, on the other hand, follow the notion of naturalistic selection to achieve optimum outcomes by choosing the preferable fit data to cope with unexpected circumstances [17]. Because of AI technologies advancement and the shortcomings of traditional computational methods, AI-based models are now being integrated into nearly all areas of research fields, including linguistics, engineering, and medicine [17]. AI modeling techniques' capabilities in handling multi-dimensional and noisy information prove the rise in AI applications fields. AI has been widely applied in environmental engineering to solve air pollution, water, wastewater treatment modeling, soil remediation simulation, groundwater contamination, and solid waste management (SWM) strategy planning [16]. To predict concentration of contaminants and particulate issue, AI-based prohibition managements technicalities such as ANN, Adaptive Neuro-Fuzzy Inference System (ANFIS), and multilayer perception (MLP) models have been applied [18, 19]. MLP is a vital modeling algorithm for calculating atmospheric levels for nitrogen dioxide, carbon monoxide, and ozone [20]. ANFIS, on the other hand, has been useful for predicting and optimizing processes for water and wastewater treatment plants [21, 22]. ANFIS has efficiently forecast the production of methane and volatiles solids flowing into the anaerobic digester in the wastewater treatment facility and coagulation dose optimization for turbidity removal in the water treatment plant [23]. AI is currently used extensively in SWM to forecast waste generation trends, improve routes for waste collection vehicles, defining waste treatment facilities, and model processes for diversion of waste, among other things. A few AI research researches covered particular fields of implementation related to waste, like medical waste management modeling and optimization, waste incineration, and generation of biogas [17, 24, 25]. The numerous Useful AI mechanisms to modeling the weight, structure, and average generation of waste were discussed [16]. Other analysis researches focused exclusively on defining the AI techniques used to forecast MSW generation rate based on sociodemographic and economic factors [26, 27]. Finally, in de Souza Melaré [28], AI-based optimization approaches in SWM were addressed to prediction waste production, control systems for collecting wastes, monitoring waste containers, and identifying disposal location. It is fine that there is a shortage of a review article that brings all the research work carried out in the various SWM fields on AI applications. A systematic critical survey examining AI-based techniques in different SWM processing has not been performed to date.

To inspect the possible use of AI techniques to resolve the broad range of problems affecting SWM, a detailed and comprehensive discussion on the current work and the results reported is crucial to promoting further developments in this area. This chapter provides a comprehensive SLR and thorough discussion of AI techniques that are have being used in their various phases, from compilation to final disposal, to enhance existing SWM schemes. Hybrid AI-based frameworks and comprehensive data with AI/non-AI frameworks are thoroughly explored to comprehend the methodology comprehensively. The possible obstacles, drawbacks, and possibilities for study are discussed. This chapter provides a methodology literature survey and an in-profound discussion of AI applications that have the opportunity to equip modern, ecologically benign systems, and economical for prospective SWM strategies, like more researchers, are interested in AI models in SWM. The chapter's goal is to direct SWM investigators concerned with AI models from their perspective areas of studies out of the key side of research, including AI models, their benefits and drawbacks, performance, and software platforms. The primary AI mechanism described in the surveys are evaluated and debated in the corresponding department; those include hybrid and individual AI framework. This section also addresses studies that compare AI and other technologies other than AI and offers summaries of the simulation platforms specified in the survey. An in-deepness argumentation of the different SWM areas in which AI techniques have been widely used is provided in the next section. This chapter concludes with the challenges and constraints faced by the SWM application of AI models and potential future recommendations for AI research and AI developments that depend on SWM.

4   Waste Management During COVID-19

Many forms of extra medical and hazardous waste are developed during the outbreak, including contaminated masks, gloves, and other preventative equipment, along with a more significant amount of non-infected objects of the same type, refer to Fig. 2. The amount of food and plastic waste has also increased during the pandemic. According to the total population and the percentage of the urban population, the number of 102 facemasks used daily in the world is estimated to reach over 7 billion. Based on the estimation of daily use of face mask. The daily face mask was adapted from Nzediegwu and Chang [38] using Eq. (1)

$$D = P \times Up \times F^{MAR} \times \left(F^{MGP}/10,000\right) \tag{1}$$

Where

- D is the Daily face mask use (pieces)
- P is the Population numbers
- Up is the Urban population
- F^{MAR} = Face masks acceptance rate = X%
- F^{MGP} = Assumption that each person in the general population uses one face mask each day

Latest ambient propagation studies have contributed to suggestions that masks should be used in public settings. Sound administration of that waste is capable

Fig. 2 Waste based face mask

of mitigating unintended impacts on human or environmental health. Appropriate detection, collection, isolation, storage, transmission, handling and disposal, and substantially related aspects are needed, including disinfection, staff safety, and training, are required in order to manage biomedical and healthcare waste effectively. Figure 2 summarises approaches to the disposal of waste during the COVID-19 disease. Contaminated waste sources are confined to hospitals.

To cope with the increasingly growing of infected people, even advanced health care companions have become inadequate. Self-isolated at home, patients with mild symptoms produce polluted SWM. It includes a significant systemic reform in the management of waste, from the principles of sorting, Recycling, and waste disposal to the waste collection workers' safety procedure. It is possible to find various safety measures in ACRş. During each pandemic, public health is given priority over all other concerns. Recovery plans and packages of economic stimulants are be created. We call for renewable energy to be at the core of AI's stimulant schema for the coronavirus crisis. The plastic surge's economic and environmental effects have not so far been thoroughly examined. The quantity of waste overhang to obliterate existent remediation and recycling facilities discompose the hazard of consequential contamination of unsuitable waste management. The pandemic results should create a more robust and different future world as a foundation/lesson. A significant question is how weight management AI strategies set to cope with the undulation will transform after the pandemic into long-term waste management preference.

5 Artificial Intelligence Role in Forecasting Waste Management

The literature review suggested that ANN, Adaptive neuro-fuzzy inference systems, support vector machine (SVM) [32], k-nearest neighbors [24, 31, 37], Genetic algorithm (GA), and hybrid models are the commonly used AI systems modeling and optimizing SWM processes. The most repeated AI model was ANN; numerous ANN algorithms were used in the literature, included radiant foundation function (RBF), MLP, backpropagation (BP), feedforward, self and repeated regression ANNs [29, 30].

GA [33] was another widely used algorithm that included multivariate and multiple LR, followed by LR, Regression, in addition to gradient supports. Other models were utilized less usually in individual research, like ANFIS [34–36], wavelet transform (WT), random forest (RF), data mining, K-means, rough sets, Naïve Bayes, Q-type clustering, logistic model tree, non-inferior set estimation (NISE), ant colony optimization, Artificial Immune System (AIS), and targeted programming. It must know this is only for trainable version, namely ANFIS, of the FL models, are protected by this review; a review of FL waste management applications can be found elsewhere.

It is necessary for outlook designing for robust AI-based SWM implementation to recognize the limitations of these techniques; elementary examples of these challenges are as follows:

1. A major barrier simulating the application of AI systems is insufficient data. For training and calibration purposes, AI models are primarily guided by comprehensive data sets. The absence or incomplete waste data is also plaguing current research. This is partly because, with minimal reliable documentation and scarce sensory knowledge, SWM industries are mostly outdated, predominantly in improving countries.

2. The various AI models and their faster growth are distracting the attempts to incorporate AI into SWM. An abundant number of models were seen in the reviewed literature, each showing good results compared with traditional methods. With a few exemptions, however, there was no topic in particular areas of radical research work. In another word, considering the number of trials, overall improvement doesn't appear to be as substantial as expected.

3. A key problem in the widespread implementation of these technologies is the black box design of AI models. Most AI models are difficult to reproduce because they are primarily founded on large data sets mostly covered or unobserved in posts. This is one of the primary factors behind the deficiency of persistence in the SWM area in many AI application efforts.

4. The majority of studies have applied AI models directly to resolve specific waste problems. Customized AI solutions classifying the particular characteristics and characteristics of SWM systems have rarely been identified. It can be achieved through an in-depth joint discussion dialogue of multidisciplinary computer science and waste management teams, focusing on extremely skilled technical workers in AI.
5. Compared to conventional approaches, the SWM action organizations are increasingly embracing the transfer across AI. This transformation can be accomplished by narrowing the gap between industry and researchers, e.g., by supporting technology start-ups and collaborations among academic institutions and small and medium-sized enterprises (SEMs) or by research and development departments large SWM companies.
6. In several heavily populated nations, SWM is governed by the government-controlled municipal/city authority. Health inspectors, public health engineers, mechanical engineers, technicians and waste staff are mostly part of the SWM system. By prioritizing health and environmental problems, the conventional approach of handling waste needs to be updated. However, we can combine it with AI techniques to handle it easily.
7. During the pandemic, waste pickers are more vulnerable to health risks, and the longer the exposure to the waste, the greater the likelihood of COVID-19 being transmitted. The use of automated technology, ICT for the collection, storage, and disposal of waste, which AI techniques support, should be emphasized. Before entering the landfill site, all waste collection vehicles can go through an automatic disinfection process, and CNN techniques help select the short path to a landfill site.
8. Especially after one-time usage during the pandemic, plastics used for food containers/wrapping will most likely continue to increase. Instead of offering a second idea of reusing, people will throw away more plastic bottles/containers. To ensure an effective and sustainable waste management system, segregating biomedical waste, plastic waste, and food waste at the household level is a must. AI techniques can be given to each household guidelines on segregation and disposal of waste.

6   Conclusions

The evaluation of the various AI models used in SWM applications was the focus of this systematic analysis. A comparison was made of AI algorithms' performances and addressed the intensity and weaknesses of AI implementation in waste management. This Literature Review showed that various forms of AI models were used to forecast, model, simulate and optimize SWM systems, both stand-alone and hybrid. In general, as most of the waste management problems are fundamentally congregation and poorly described, it is clear that conventional approaches, instituted on mechanical models and rigorous algorithms, don't in many cases seem to provide a sufficient solution, in particular for those hard to get data insufficiency. AI models provide an alternate and productive solution that has drawn considerable interest in the scientific society. While research in this area is progressing speedily, SWM systems based on AI are still developing and research.

References

1. MacKenzie, D.: Covid-19 goes global. New Sci. **245**(3271), 7 (2020)
2. Worldometer. COVID-19 Coronavirus pandemic (2020). http://www.worldometersinfo/corona virus/#countries%3C. Accessed 08 April 2020
3. Rajmohan, K.V.S., Ramya, C., Viswanathan, M.R., Varjani, S.: Plastic pollutants: effective waste management for pollution control and abatement. Current Opinion Environ. Sci. Health **12**, 72–84 (2019)
4. Windfeld, E.S., Brooks, M.S.L.: Medical waste management–a review. J. Environ. Manage. **163**, 98–108 (2015)
5. Liu, H.C., You, J.X., Lu, C., Chen, Y.Z.: Evaluating healthcare waste treatment technologies using a hybrid multi-criteria decision making model. Renew. Sustain. Energy Rev. **41**, 932–942 (2015)
6. Yu, H., Sun, X., Solvang, W.D., Zhao, X.: Reverse logistics network design for effective management of medical waste in epidemic outbreaks: Insights from the coronavirus disease 2019 (COVID-19) outbreak in Wuhan (China). Inter. J. Environ. Res. Public Health **17**(5), 1770 (2020)
7. Sharma, H.B., Vanapalli, K.R., Cheela, V.S., Ranjan, V.P., Jaglan, A.K., Dubey, B., Goel, S., Bhattacharya, J.: Challenges, opportunities, and innovations for effective solid waste management during and post COVID-19 pandemic. Resour. Conserv. Recycl. **162**, (2020)
8. Mihai, F.C.: Assessment of COVID-19 waste flows during the emergency state in romania and related public health and environmental concerns. Intern. J. Environ. Res. Public Health **17**(15), 5439 (2020)
9. Ma, Y., Lin, X., Wu, A., Huang, Q., Li, X., Yan, J.: Suggested guidelines for emergency treatment of medical waste during COVID-19: Chinese experience. Waste Disp. Sustain. Energy p. 1 (2020)
10. Calma, J.: The COVID-19 pandemic is generating tons of medical waste (2020). Retrieved 27 April 2020
11. ACR +.: Municipal waste management and COVID-19 (2020). Retrieved https://buff.ly/3db vNs3
12. Sarkodie, S.A., Owusu, P.A.: Global assessment of environment, health and economic impact of the novel coronavirus (COVID-19). Environ. Develop. Sustain. 1–11 (2020)
13. W4C.: Understanding Medical Waste Management to Curb the Transmission of COVID-19 (2020). Retrieved https://buff.ly/30PMy
14. Weforum.: The plastic pandemic is only getting worse during COVID-19 (2020). Retrieved https://buff.ly/2FoGNXn
15. AMSA.: Waste management and cleaning services in Milan during COVID-19 (2020). Retrieved https://buff.ly/2TKS1JU
16. Yetilmezsoy, K., Ozkaya, B., Cakmakci, M.: Artificial intelligence-based prediction models for environmental engineering. Neural Netw. World **21**(3), 193 (2011)
17. Kalogirou, S.A.: Use of genetic algorithms for the optimal design of flat plate solar collectors (2003)
18. Roy, S.: Prediction of particulate matter concentrations using artificial neural network. Resour. Environ **2**(2), 30–36 (2012)

19. Shu, H.Y., Lu, H.C., Fan, H.J., Chang, M.C., Chen, J.C.: Prediction for energy content of Taiwan municipal solid waste using multilayer perceptron neural networks. J. Air Waste Manag. Assoc. **56**(6), 852–858 (2006)
20. Agirre-Basurko, E., Ibarra-Berastegi, G., Madariaga, I.: Regression and multilayer perceptron-based models to forecast hourly O3 and NO2 levels in the Bilbao area. Environ. Model Softw. **21**(4), 430–446 (2006)
21. Cakmakci, M.: Adaptive neuro-fuzzy modelling of anaerobic digestion of primary sedimentation sludge. Bioprocess Biosyst. Eng. **30**(5), 349–357 (2007)
22. Chun, M.G., Kwak, K.C., Ryu, J.W.: Application of ANFIS for coagulant dosing process in a water purification plant. In FUZZ-IEEE'99. 1999 IEEE International Fuzzy Systems. Conference Proceedings (Cat. No. 99CH36315). IEEE, vol. 3, pp. 1743–1748 (1999)
23. Niska, H., Serkkola, A.: Data analytics approach to create waste generation profiles for waste management and collection. Waste Manag. **77**, 477–485 (2018)
24. Enitan, A.M., Adeyemo, J., Swalaha, F.M., Kumari, S., Bux, F.: Optimization of biogas generation using anaerobic digestion models and computational intelligence approaches. Rev. Chem. Eng. **33**(3), 309–335 (2017)
25. Qin, X.S., Huang, G.H., He, L.: Simulation and optimization technologies for petroleum waste management and remediation process control. J. Environ. Manage. **90**(1), 54–76 (2009)
26. Goel, S., Ranjan, V.P., Bardhan, B., Hazra, T.: Forecasting solid waste generation rates. In: Modelling Trends in Solid and Hazardous Waste Management, pp. 35–64. Springer, Singapore (2017)
27. Kolekar, K.A., Hazra, T., Chakrabarty, S.N.: A review on prediction of municipal solid waste generation models. Proced. Environ. Sci. **35**, 238–244 (2016)
28. de Souza Melaré, A.V., González, S.M., Faceli, K., Casadei, V.: Technologies and decision support systems to aid solid-waste management: a systematic review. Waste Manag. **59**, 567–584 (2017)
29. Duda, R.O., Hart, P.E., Stork, D.G.: Pattern classification. John Wiley & Sons (2012)
30. Ozkaya, B., Demir, A., Bilgili, M.S.: Neural network prediction model for the methane fraction in biogas from field-scale landfill bioreactors. Environ. Model Softw. **22**(6), 815–822 (2007)
31. Chen, Z., Wang, L., Wu, W., Jiang, Z., Li, H.: Monitoring plastic-mulched farmland by Landsat-8 OLI imagery using spectral and textural features. Remote Sens. **8**(4), 353 (2016)
32. Harrington, P., 2012. Machine learning in action. Manning Publications Co
33. Meyer-Baese, A., Schmid, V.J.: Pattern recognition and signal analysis in medical imaging. Elsevier (2014)
34. Tiwari, M.K., Bajpai, S., Dewangan, U.K.: Prediction of industrial solid waste with ANFIS model and its comparison with ANN model-A case study of Durg-Bhilai twin city India. Inter. J. Eng. Innov. Technol. (IJEIT) **6**(2), 192–201 (2012)
35. Chen, H.W., Chang, N.B.: Prediction analysis of solid waste generation based on grey fuzzy dynamic modeling. Resour. Conserv. Recycl. **29**(1–2), 1–18 (2000)
36. Noori, R., Abdoli, M.A., Ghazizade, M.J., Samieifard, R.: Comparison of neural network and principal component-regression analysis to predict the solid waste generation in Tehran. Iranian J. Public Health 74–84 (2009)
37. Meade, N.: A comparison of the accuracy of short term foreign exchange forecasting methods. Int. J. Forecast. **18**(1), 67–83 (2002)
38. Sangkham, S.: Face mask and medical waste disposal during the novel COVID-19 pandemic in Asia. Case Stud. Chem. Environ. Eng. **2**, 100052 (2020). https://doi.org/10.1016/j.cscee.2020.100052

Carbon Monoxide Air Pollution Monitoring Approach in Africa During COVID-19 Pandemic

Reham Gharbia⑩ and Aboul Ella Hassanien

Abstract In December of 2019, a new coronavirus, called SARS-CoV-2, began in Wuhan, China, known as COVID-19. The COVID-19 epidemic has expanded dramatically. The epidemic lockdown of COVID-19 affected air quality due to severe changes in human behavior. Recent research shows that air quality has improved during the COVID-19 pandemic due to social lockdown measures. Africa can avoid outbreaks of respiratory diseases, including COVID-19, with cleaner air. However, monitoring air pollution over the continent has required complex efforts to track air quality due to policies, borders, and the continent's nature. Now, satellite data can monitor air quality in cities and countries and bridge the gap in African countries without ground sensors. This paper examines the impact of lockdown on COVID-19 on air pollution in Africa. This study included measurements of the total column concentration of carbon monoxide (CO) from the tropospheric monitoring instrument (TROPOMI) on the Sentinel-5 precursor satellite. This study analyzed the global carbon dioxide concentration in Africa before, during, and after the COVID-19 pandemic from 15 March to 6 June 2019 and 2020. The results here indicate a small decrease in the CO levels because more than nine hundred million people in Africa depend on energy sources polluted for household uses, resulting from wood-burning, kerosene, and charcoal. Many millions of people in Africa live in small crowded houses, bad ventilation and building materials. These dangers cause rising air pollution.

Keywords COVID-19 · Carbon monoxide · Africa continent · Sentinel-5 precursor (S-5P) platform · Air quality monitoring · Remote sensing

R. Gharbia (✉)
Nuclear Materials Authority, Cairo, Egypt
URL: http://www.egyptscience.net

A. E. Hassanien
Faculty of Computer and Artificial Intelligence, Cairo University, Cairo, Egypt

R. Gharbia · A. E. Hassanien
Scientific Research Group in Egypt (SRGE), Cairo, Egypt

© The Author(s), under exclusive license to Springer Nature Switzerland AG 2021
A. E. Hassanien et al. (eds.), *The Global Environmental Effects During and Beyond COVID-19*, Studies in Systems, Decision and Control 369,
https://doi.org/10.1007/978-3-030-72933-2_6

1   Introduction

In December of 2019, a novel coronavirus appeared as SARS-CoV-2, started in Wuhan, China, and known as the COVID-19 [1]. The total number of death from COVID-19 has approximately about million and twenty-nine million confirmed cases. To stop the disease's spread, we must know why some places have higher numbers of confirmed cases and deaths than others better. COVID-19 is induced by hard acute respiratory symptoms, coronavirus 2 (SARS-CoV-2) [2–4]. Usually, most SARS-CoV-2 infected cases have moderate symptoms, including dry cough, fever, and soreness [5]. However, some cases could have hard and even deadly complications as Acute Respiratory Distress Syndrome [6]. To control the spread of COVID-19, during the first half of 2020, most of the world's countries have forced rigorous procedures such as closing factories, businesses, restricting travel, and issuing stay-at-home rules. The lockdown procedures have caused the spread of the virus [7] and produced large reductions in international demand for fossil fuels [8]. In the last decades, researchers have searched to conclude how industrial activity and transportation affect air quality, especially in cities, where emissions from internal combustion engine vehicles constitute the primary source of air pollution. COVID-19 restrictions turn into a natural experiment that lets researchers connect air pollution changes to reductions in use and formulate transportation policies to improve air quality [9].

Exposure to air pollution is responsible for about 700,000 deaths in Africa per year. More than 14% of all non-communicable diseases Occur because of air pollution more than the other cause of factors, including alcohol, high sodium intake, and a strict diet [10]. Furthermore, the NCDs which are pre-existing conditions that raise the risk of death for COVID-19. It correlates with the increase in exposure to air pollution [11].

Carbon monoxide (CO) is a widespread environmental gaseous pollutant whose focus indicates air quality standards in different countries or regions. And the incomplete oxidation of fossil fuels is the main source of ambient carbon dioxide, contributing to N50% of emissions in urban areas. Other sources such as industrial and natural processes are less clear [12].

Carbon monoxide binds avidly to hemoglobin, which causes hypoxemia and reduces oxygen delivered to tissues at high concentrations [13, 14]. Carbon monoxide is universally known as the silent killer due to its ability to bind hemoglobin strongly more than oxygen, causing the probability of asphyxia-related deaths at exposure to high levels of hypoxic damage to tissues at exposure to lower concentrations [15]. Plenty of epidemiological studies have proven that short-term exposure to ambient Carbon monoxide is joined with raised risks of adverse health effects, mainly all-cause mortality and cardiovascular diseases [16].

The population restriction is commanding extreme energy use changes, with expected CO emissions impacts [17]. The levels of Carbon monoxide (CO) emissions decreased by an approximated 8.6%, nitrogen dioxide (NO2) and other air pollutants have also decreased across over the world [18, 19] depended on pointers of energy usage between January and April of 2020 as opposed to the same period in the last year [17]. The reductions in air pollution and CO emissions may be short-lived. The global CO emissions have ricocheted and returned to their earlier rates [20]. Indeed, economic activities are already retuning in many countries. Governments have already announced or suggested large economic activity packages to encourage recovery [21] without any climate change reduction co-objective to protect the environment.

Recently, satellite devices' abilities for sensing the troposphere have developed and presented the way for monitoring and a better vision of atmospheric pollution processes and emissions. Satellite instruments implement global measures of many pollutants such as ozone, CO, NO2, and aerosols. Satellite platforms have the characteristic of providing measurements with global coverage but at a moderately low temporal resolution. Geostationary Earth orbit satellite platforms produce observations at a continental scale, i.e., not global, but at a much more powerful temporal resolution [22, 23]. Copernicus is the current European program for the ability of earth observation. The Copernicus Atmospheric's chief objective is to provide information on atmospheric climate changes to support sustainable development and the global protection of the environment.

The Sentinel-5 Precursor (S-5P) platform data for CO total column was used over Africa during the COVID 19 pandemic lockdown in a recent chapter. The main objective is to investigate the CO during and after the lockdown on Africa. The rest of this paper is described as follows. Section 2 presents the proposed air pollution monitoring modeling. Section 3 illustrates the results and discussion, and finally, Sect. 4 include the conclusion.

2   The Proposed Air Pollution Monitoring Modeling

2.1   Data Collection

Evaluation of air quality over Africa during the COVID-19 lockdown of 2019–2020 was conducted by comparing CO concentrations with the same days of the previous period of 2019–2020, see Fig. 1. The Tropospheric Monitoring Instrument (TROPOMI) was launched on the Copernicus Sentinel-5 Precursor satellite on 13 October 2017. On 11 July 2018, the first data was released, and it is providing a vast improvement in accuracy for monitoring air pollutants. These maps display trace gases like carbon monoxide, ozone, nitrogen dioxide, and methane. Extra data produces like the Aerosol index can present atmospheric volcanic ash for flying protection and high UV radiation levels. High hopes are set on Sentinel 5P, as it provides the Tropomi instrument - the various advanced multispectral imaging spectrometer up to date. The instrument measures wave in the visible (VIS), near (NIR), ultraviolet (UV), and short-wavelength infrared (SWIR). Because of its very high resolution of up to 7×3.5 km, the Tropospheric Monitoring Instrument (Tropomi) can measure the air pollution over cities. Initial data demonstrated the hot spots of

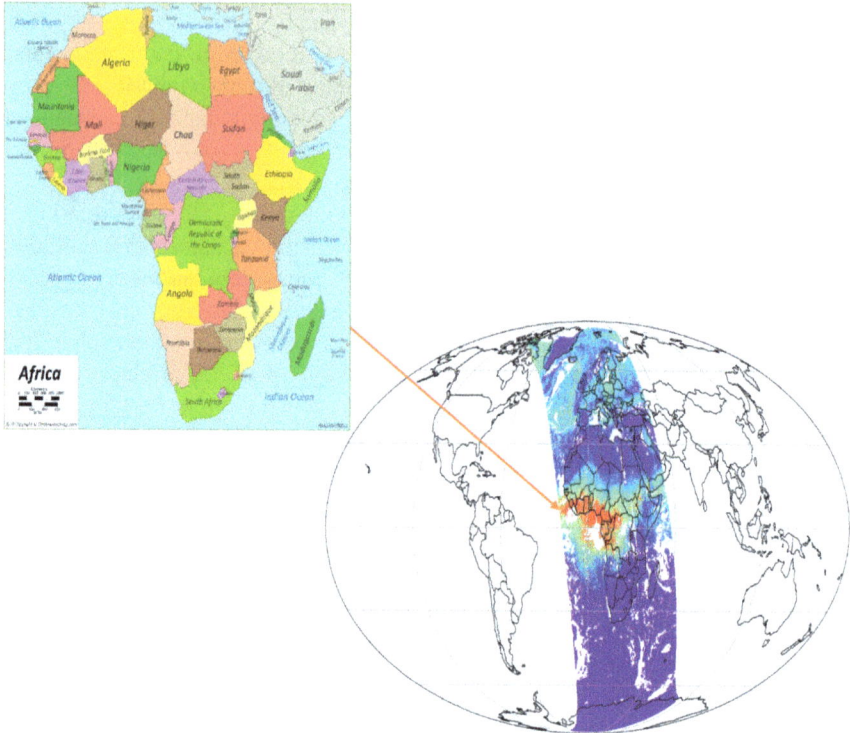

Fig. 1 The CO concentration map over Africa on 15 March 2020

nitrogen dioxide pollution over Europe, the Middle East, and Africa. Sentinel's swath width of 2600 km allows it to map all the planets every 24 h. The Sentinel 5 Precursor mission targets providing information and monitoring the air quality and climate in the timeframe 2017–2023. The TROPOMI will provide daily measure observations of atmospheric constituents, including nitrogen dioxide, carbon monoxide, ozone, sulfur dioxide, methane, formaldehyde, and cloud and aerosol properties.

The CO Satellite Measurements dataset produces global daily averages of CO density in the total atmospheric column at a 5.5×3.5 km (km) resolution. Each value recorded in the dataset provides the density of CO found between Earth's surface and the atmosphere's top. The CO density is given with units of moles of CO per square meter of air (mol/m^2).

2.2 Carbon Monoxide Monitoring Modeling

This study investigates the relationship between the COVID-19 lockdown and daily air pollutants of carbon monoxide in Africa. The proposed air pollution monitoring modeling is described in Fig. 2, and the main steps are described as follows:

- Load and browse Sentinel-5P data: Open one individual Sentinel-5P netCDF file with NetCDF4 library in python. The dataset object contains information about the general data structure of the dataset. The variables of Sentinel-5P data are organized in groups, which is analogous to directories in a file system
- Merge files: Merge more than one file to obtain Africa's file

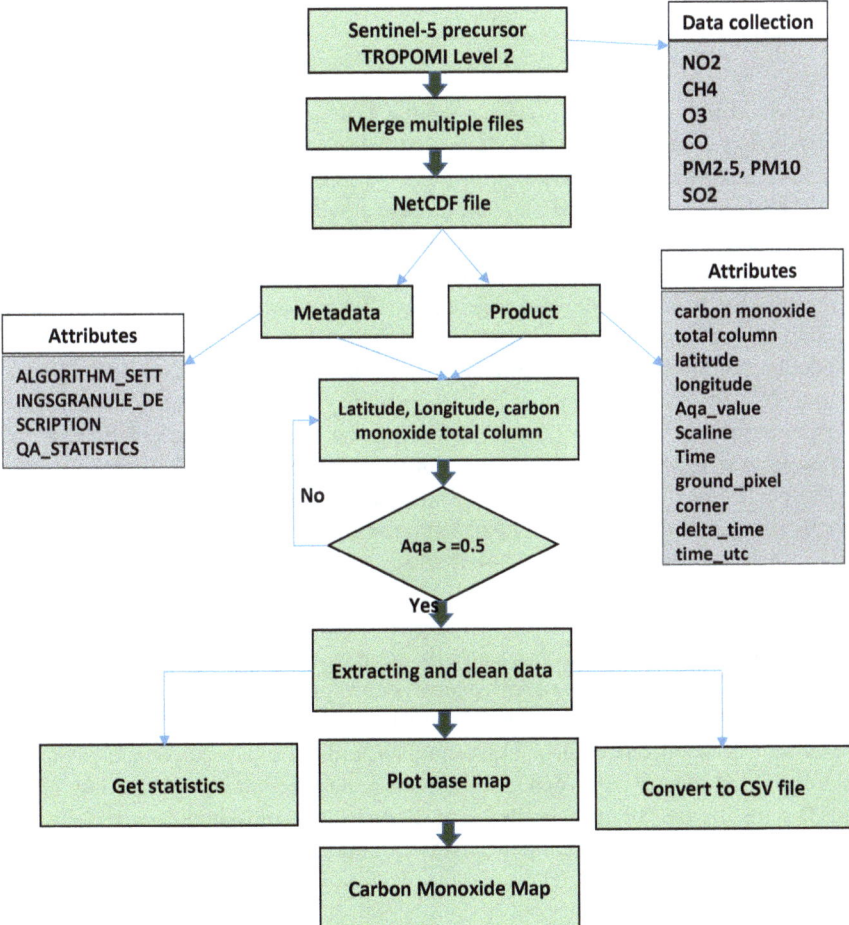

Fig. 2 The proposed air pollution minoring modeling

- Specify one variable of interest and get more detailed information about the variable. For example, carbonmonoxide_total_column is the atmosphere mole content of carbon monoxide, has the unit mol m-2, and is a 3D variable. You can do this for the available variables but also the dimensions of latitude and longitude.
- Extract clean data: Retrieve the clean data, which is Aqa statistic more than 0.5 value
- Plot the processing Sentinel-5P data for Africa
- Retrieve and visualize Sentinel-5P data
- Calculate the statistic parameters such as average, Standard deviation and median, etc.
- Convert the clean data to a CVS file.

3 Results

We collected air pollution concentrations in the Africa continent for carbon monoxide for ten weeks from 15 March to 6 June. 2019 (left) and the period from 15 March to 6 June 2020 (right), as indicated in Fig. 3. To analyze the relation between the COVID-19 pandemic lockdown and the daily average carbon monoxide in the Africa continent before, during, and post the lockdown.

All images show a wide range of CO levels across middle Africa, from near zero in the west to above 40 in the east. All images offer a wide range of CO levels across middle Africa, from close zero in the west to above 40 in the east, reflecting the rise in population, human and industrial activity from the west to east in the continent. All images show a wide range of CO levels across middle Africa, from near zero in the west to above 40 in the east, reflecting the rise in population, human and industrial activity from the west to east in the continent. Table 1 summarizes the CO levels before and during the COVID-19 pandemic lockdown in Africa and the relative change in the period from 15 March to 6 June in 2019 and 2020.

The data presented here (Fig. 4) indicate a small decrease in the CO levels comparing with the European continent [24, 25]. More than nine hundred million people in Africa depend on polluting energy sources for household uses, and about 600 million are without having electricity sources (International Energy Agency 2019). The millions of people in Africa have been exposed to high air pollution levels during the COVID-19 period because of the exposure to high CO levels resulting from the burning of wood, charcoal, and kerosene. These dangers are exacerbated by many millions living in small crowded spaces, inadequate ventilation, and building materials that raising the pollution burden [26]. Furthermore, most houses will use whatever available fuel: scrap tires, plastic waste, cloth rags, and other unconventional materials [27].

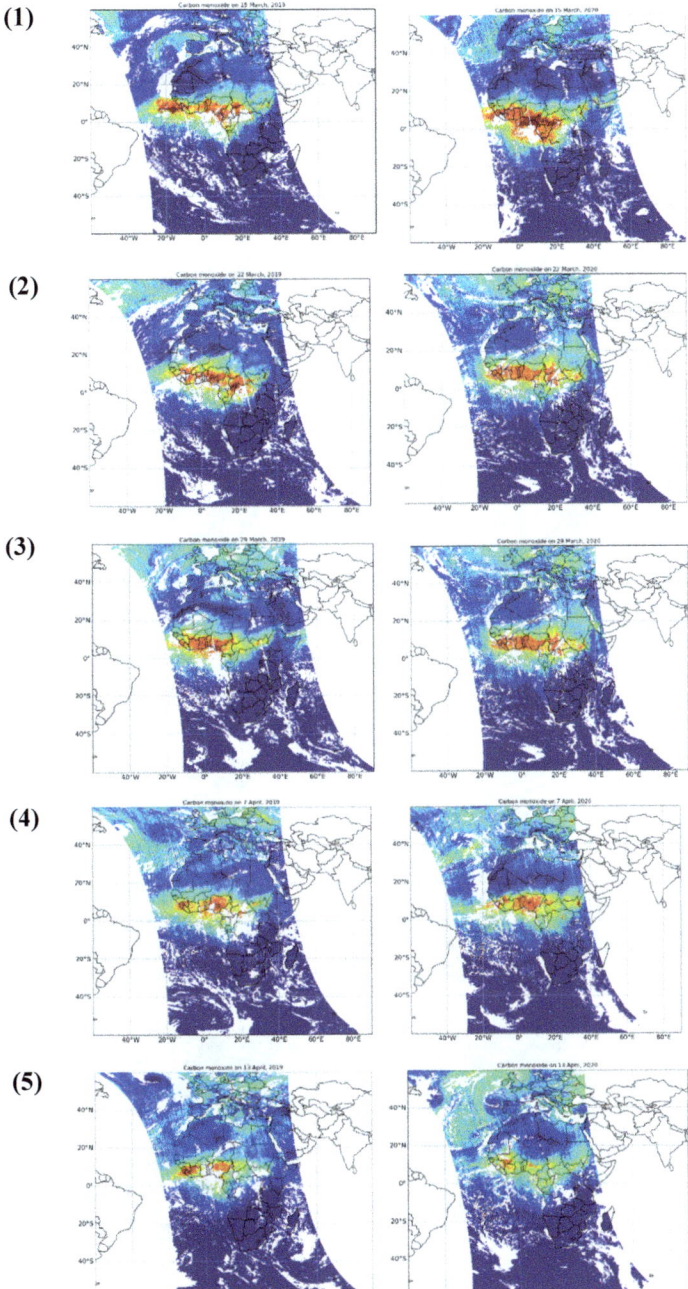

Fig. 3 The average carbon monoxide's spatial distribution for ten weeks in Africa before and during the lockdown

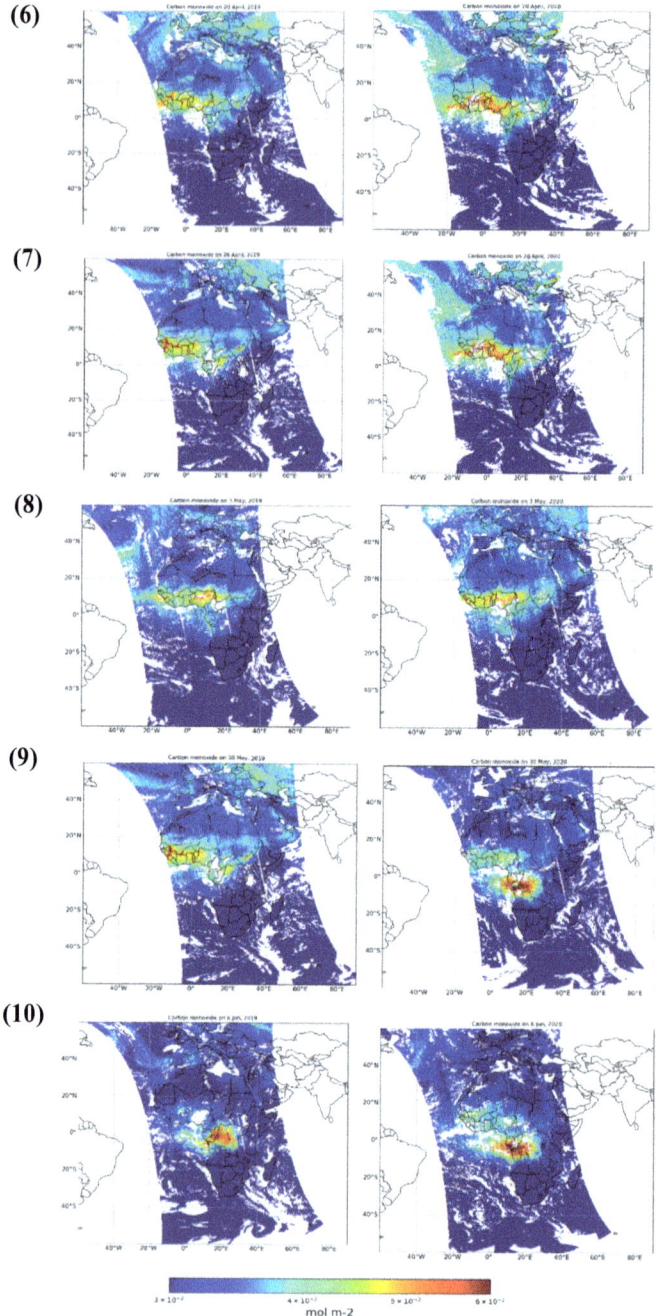

Fig. 3 (continued)

Table 1 CO levels before and during the COVID-19 pandemic lockdown in Africa

	15-Mar	22-Mar	29-Mar	7-Apr	13-Apr	20-Apr	26-Apr	3-May	30-May	6-Jun
2019	0.011	0.011	0.012	0.01	0.01	0.009	0.009	0.008	0.009	0.009
2020	0.011	0.012	0.011	0.011	0.01	0.01	0.009	0.008	0.006	0.007
Relative change	0	−0.001	0.001	−0.001	0	−0.001	0	0	0.003	0.002

Fig. 4 The daily average of the CO in the period 15 March to 6Jun 2019 and 2020

4   Conclusion and Future Work

In this paper, the lockdown's potential effects due to the COVID-19 pandemic on the CO levels were investigated in Africa. The CO levels from 15 March to 6 June 2019 compared with the same period during the COVID-19 pandemic lockdown in 2020. the results provide a small decrease in the air pollution of CO in Africa the resulting poor household uses, resulting from wood-burning, kerosene, and charcoal. Many millions of people in Africa live in small crowded houses, lousy ventilation and building materials. These dangers cause rising air pollution.

A proposal to monitor and analyze the global concentration of air pollution emissions of satellites based on machine learning may be considered in future work. The idea is to combine satellite imagery of smoke being generated from power plants, factories, and other sources with sensors for infrared and thermal imaging and air pollution, sensors such as CO, NO^2, and CH^4 sensor data, and then analyze the results via machine learning for improving emissions data. This approach will provide much more accurate and up-to-date information than the approaches widely available to governments and researchers.

References

1. Lu, H., Stratton, C.W., Tang, Y.W.: Outbreak of pneumonia of unknown etiology in Wuhan, China: The mystery and the miracle. J. Med. Virol. **92**(4), 401–402 (2020)
2. Dong, E., Du, H., Gardner, L.: An interactive web-based dashboard to track COVID-19 in real time. Lancet. Infect. Dis **20**(5), 533–534 (2020)
3. Sohrabi, C., Alsafi, Z., O'Neill, N., Khan, M., Kerwan, A., Al-Jabir, A., Iosifidis, C., Agha, R.: World Health Organization declares global emergency: a review of the 2019 novel coronavirus (COVID-19). Inter. J. Surg. (2020)
4. Zhou, F., Yu, T., Du, R., Fan, G., Liu, Y., Liu, Z., Xiang, J., Wang, Y., Song, B., Gu, X., Guan, L.: Clinical course and risk factors for mortality of adult inpatients with COVID-19 in Wuhan, China: a retrospective cohort study. Lancet (2020)
5. Huang, C., Wang, Y., Li, X., Ren, L., Zhao, J., Hu, Y., Zhang, L., Fan, G., Xu, J., Gu, X., Cheng, Z.: Clinical features of patients infected with 2019 novel coronavirus in Wuhan China. lancet **395**(10223), 497–506 (2020)
6. Chen, H., Guo, J., Wang, C., Luo, F., Yu, X., Zhang, W., Li, J., Zhao, D., Xu, D., Gong, Q., Liao, J.: Clinical characteristics and intrauterine vertical transmission potential of COVID-19 infection in nine pregnant women: a retrospective review of medical records. Lancet **395**(10226), 809–815 (2020)
7. Tian, H., Liu, Y., Li, Y., Wu, C.H., Chen, B., Kraemer, M.U., Li, B., Cai, J., Xu, B., Yang, Q., Wang, B.: An investigation of transmission control measures during the first 50 days of the COVID-19 epidemic in China. Science **368**(6491), 638–642 (2020)
8. Liu, F., Page, A., Strode, S.A., Yoshida, Y., Choi, S., Zheng, B., Lamsal, L.N., Li, C., Krotkov, N.A., Eskes, H., Veefkind, P.: Abrupt decline in tropospheric nitrogen dioxide over China after the outbreak of COVID-19. Sci. Adv. 2992 (2020)
9. Baldasano, J.M.: COVID-19 lockdown effects on air quality by NO2 in the cities of Barcelona and Madrid (Spain). Sci. Total Environ. **741**, (2020)
10. GBD.: Global burden of Disease Study 2015-Results by risk factor-country level (online data base-Viz Hub-GBD Compare). Seattle (2017). https://vizhub.healthdata.org/gbdcompare
11. Mbandi, A.M.: Air Pollution in Africa in the time of COVID-19: the air we breathe indoors and outdoors. Clean Air J. **30**(1), 1–3 (2020)
12. Janssens-Maenhout, G., Crippa, M., Guizzardi, D., Muntean, M., Schaaf, E., Dentener, F., Bergamaschi, P., Pagliari, V., Olivier, J.G., Peters, J.A., Van Aardenne, J.A.: EDGAR v4. 3.2 Global Atlas of the three major greenhouse gas emissions for the period 1970–2012. Earth Syst. Sci. Data **11**(3), 959–1002 (2019)
13. McKee, M., Stuckler, D.: If the world fails to protect the economy, COVID-19 will damage health not just now but also in the future. Nat. Med. **26**(5), 640–642 (2020)
14. Wu, L., Wang, R.: Carbon monoxide: endogenous production, physiological functions, and pharmacological applications. Pharmacol. Rev. **57**(4), 585–630 (2005)
15. Xu, C., Fan, Y.N., Liang, Z., Xiao, S.H., Huang, L., Kan, H.D., Chen, R.J., Liu, X.L., Yao, C.Y., Luo, G., Zhang, Y.: Unexpected association between increased levels of ambient carbon monoxide and reduced daily outpatient visits for vaginitis: a hospital-based study. Sci. Total Environ. 137923 (2020)
16. Zhao, Y., Hu, J., Tan, Z., Liu, T., Zeng, W., Li, X., Huang, C., Wang, S., Huang, Z., Ma, W.: Ambient carbon monoxide and increased risk of daily hospital outpatient visits for respiratory diseases in Dongguan, China. Sci. Total Environ. **668**, 254–260 (2019)
17. Le Quéré, C., Jackson, R.B., Jones, M.W., Smith, A.J., Abernethy, S., Andrew, R.M., De-Gol, A.J., Willis, D.R., Shan, Y., Canadell, J.G., Friedlingstein, P.: Temporary reduction in daily global CO_2 emissions during the COVID-19 forced confinement. Nature Clim. Change 1–7 (2020)
18. Venter, Z.S., Aunan, K., Chowdhury, S., Lelieveld, J.: COVID-19 lockdowns cause global air pollution declines with implications for public health risk. MedRxiv (2020)

19. Bauwens, M., Compernolle, S., Stavrakou, T., Müller, J.F., Van Gent, J., Eskes, H., Levelt, P.F., van der A, R., Veefkind, J.P., Vlietinck, J., Yu, H.: Impact of coronavirus outbreak on NO2 pollution assessed using TROPOMI and OMI observations. Geophys. Res. Lett. **47**(11), e2020GL087978 (2020)
20. Chambers, C.A., Hopkins, R.O., Weaver, L.K., Key, C.: Cognitive and affective outcomes of more severe compared to less severe carbon monoxide poisoning. Brain Inj. **22**(5), 387–395 (2008)
21. Liu, C., Yin, P., Chen, R., Meng, X., Wang, L., Niu, Y., Lin, Z., Liu, Y., Liu, J., Qi, J., You, J.: Ambient carbon monoxide and cardiovascular mortality: a nationwide time-series analysis in 272 cities in China. The Lancet Planetary Health **2**(1), e12–e18 (2018)
22. Streets, D.G., Canty, T., Carmichael, G.R., de Foy, B., Dickerson, R.R., Duncan, B.N., Edwards, D.P., Haynes, J.A., Henze, D.K., Houyoux, M.R., Jacob, D.J.: Emissions estimation from satellite retrievals: A review of current capability. Atmos. Environ. **77**, 1011–1042 (2013)
23. Abida, R., Attié, J.L., Amraoui, L.E., Ricaud, P., Lahoz, W., Eskes, H., Segers, A., Curier, L., Haan, J.D., Kujanpää, J., Nijhuis, A.O.: Impact of spaceborne carbon monoxide observations from the S-5P platform on tropospheric composition analyses and forecasts. Atmos. Chem. Phys. **17**(2), 1081–1103 (2017)
24. Nichol, J.E., Bilal, M., Ali, M., Qiu, Z.: Air pollution scenario over China during COVID-19. Remote Sensing **12**(13), 2100 (2020)
25. Monks, P.: Coronavirus: lockdown's effect on air pollution provides rare glimpse of low-carbon future. The Conversation (2020). https://theconversation.com/coronavirus-lockdowns-effect-on-air-pollution-provides-rare-glimpse-of-lowcarbon-future-134685
26. Corburn, J., Vlahov, D., Mberu, B., Riley, L., Caiaffa, W.T., Rashid, S.F., Ko, A., Patel, S., Jukur, S., Martínez-Herrera, E., Jayasinghe, S.: Slum health: arresting COVID-19 and improving well-being in urban informal settlements. J. Urban Health 1–10 (2020)
27. Muindi, K., Kimani-Murage, E., Egondi, T., Rocklov, J., Ng, N.: Household air pollution: sources and exposure levels to fine particulate matter in Nairobi slums. Toxics **4**(3), 12 (2016)

Applications of Deep Learning in Predicting Natural Disasters Concurrent with the COVID-19 Pandemic: Short Review and Recommendations

Dalia Ezzat⬤, Sara Abdelghafar, and Aboul Ella Hassanien

Abstract Coronavirus COVID-19 is a global pandemic caused by a newly discovered Coronavirus recently found in Wuhan, Hubei Province of China, and then widespread outbreak worldwide. As the world struggles to slow the spread of COVID-19, some natural disasters coincided with COVID-19 pandemic impacts in many regions of the world. In general, the coincidence of natural disasters with pandemics leads to a worsening of the crisis in terms of the spread of the pandemic and the number of injuries and facing the disaster and the required urgent exceptional procedures. Therefore, there is a great need for predictive intelligence models that forecast the time, place, and magnitude of the disaster. The natural disaster predictive model's inclusion could enhance the strategies approach to facing the disaster by raising the degree of preparedness that reduces risks and prepares with the necessary resources—implementing the procedures that increase the degree of preventive measures to combat the pandemic spread before the occurrence of predicted natural disasters. This chapter adopts a statistical analysis approach to present some of the natural disasters concurrent with the COVID-19 pandemic and their negative effects on the increase in infection with the COVID-19. Besides the statistical analysis approach, the literature review of Deep Learning (DL) applications in natural disaster prediction is presented. The chapter concludes with recommendations that allow intensifying contingency planning procedures by using Artificial Intelligence (AI) applications that can significantly predict natural disasters. Our recommendations may also help the governments develop strategies for risk reduction of natural disasters and reduce the spread of COVID-19.

D. Ezzat (✉) · A. E. Hassanien
Faculty of Computers and Artificial Intelligence, Cairo University, Giza, Egypt
URL: http://www.egyptscience.net

S. Abdelghafar
Computer Science Department, Faculty of Science, Al Azhar University, Giza, Egypt

D. Ezzat · S. Abdelghafar · A. E. Hassanien
Scientific Research Group in Egypt (SRGE), Cairo, Egypt

© The Author(s), under exclusive license to Springer Nature Switzerland AG 2021
A. E. Hassanien et al. (eds.), *The Global Environmental Effects During and Beyond COVID-19*, Studies in Systems, Decision and Control 369,
https://doi.org/10.1007/978-3-030-72933-2_7

Keywords Artificial intelligence · Deep learning · Machine learning · COVID-19 pandemic · Statistical analysis · Natural disasters · Flood forecasting · Earthquake prediction

1 Introduction

During the recent few months, the world suffered severe human lives losses by the frightening outbreak and massive spread of the contagious pandemic Coronavirus (COVID-19). According to the daily report World Health Organization (WHO) at the end of November 2020, there are more than 65 million confirmed cases and more than 1 million and 500 thousand deaths of the COVID-19 in 220 countries, areas, or territories cases [1]. Figure 1 shows the growth of COVID-19 confirmed cases and deaths during the few past months [2].

During the world's strenuous efforts in the search for the spread limiting ways of emerging Covid-19, it was found that social distancing is one of the best ways that had a major contribution for limiting the spread that most countries of the world resorted to during the pandemic. However, social distancing and preventive measures against epidemics are vanishing by several factors. One of the most important of these factors is natural disasters. The coincidence of both a pandemic and a natural disaster exacerbates the crisis on both sides. For example, natural disasters like floods and earthquakes lead to exceptional procedures such as evacuation and sheltering into communal environments, impacting social distancing and increasing pandemic exposure risks.

On the other hand, the epidemic poses a real danger to the humanitarian responses required with natural disasters, like providing resources and medical assistance and economic, infrastructural capacity. This chapter uses both the statistical analysis and literature review approaches to present the negative impacts of natural disasters' coincidence with the COVID-19 pandemic and to review the Deep Learning (DL) applications in natural disaster prediction. Prediction of natural disasters is the set of tasks that include forecasting of attributes, time, prone area, and magnitude of the disaster [3].

The contributions of this chapter can be summarized in the following points:

1. The chapter uses the statistical analysis approach to present the impacts of the coincidence of COVID-19 with natural disasters, specifically, geophysical, hydrological, meteorological, and climatological disasters. The study is analyzed from the number of natural disasters worldwide in 2019, the number of deaths recorded as a result of natural disasters in2018 and 2019, and the number of daily confirmed cases of COVID-19 in the regions that suffered from natural disasters concurrent with the pandemic.
2. The chapter uses a literature review to present the studies related to the use of DL applications in the early prediction of natural disasters, especially earthquakes and floods that were the most frequent and deadly occurrences in conjunction

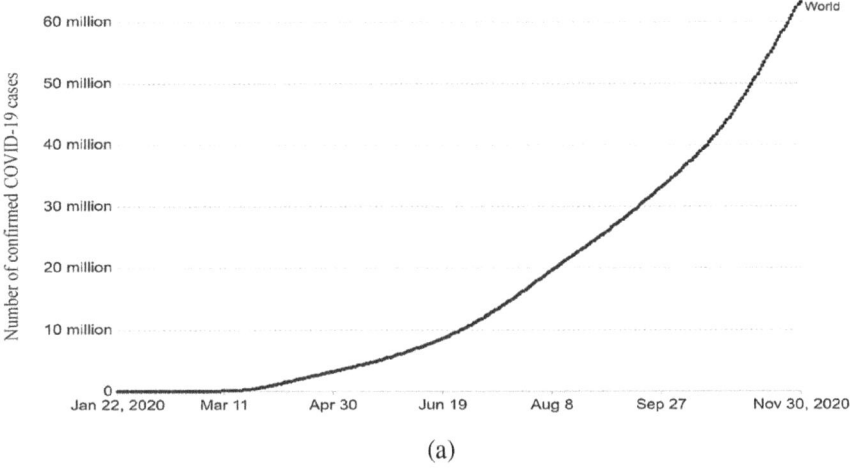

(a)

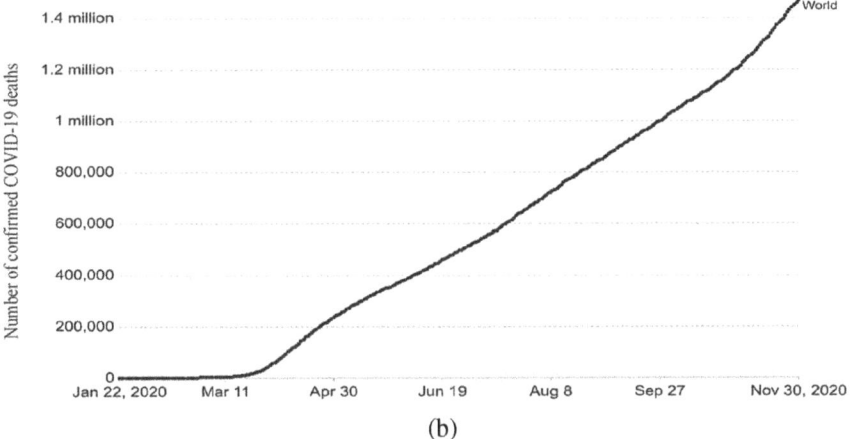

(b)

Fig. 1 The total number of COVID-19 infections and deaths from January 22, 2020 through the end of November 2020. **a** Cumulative confirmed COVID-19 cases, **b** Cumulative confirmed COVID-19 deaths. Data source [2]

with the COVID-19 pandemic. The review is limited to the studies that were proposed in 2019 and 2020.

3.   The chapter concludes with recommendations to develop strategies and planning procedures using Artificial Intelligence (AI) applications to reduce the risks of natural disasters concurrent with the COVID-19 pandemic.

The remaining sections of this chapter are organized as follows, the basics and background of natural disasters and DL architectures are described in Sect. 2. Section 3 presents the natural disasters concurrent with the COVID-19 pandemic. The review of DL applications for early natural disaster prediction is presented in Sect. 4. The recommendations are outlined in Sect. 5. Finally, conclusions and future work are given in Sect. 6.

2 Basics and Background

2.1 Natural Disasters

A natural disaster is an event that causes significant damage and is caused by the natural processes of the earth. Natural disasters can be classified into five types: geophysical, hydrological, meteorological, climatological, and biological disasters, as shown in Fig. 2. Geophysical disasters are events that arise from solid earth, such as earthquakes and volcanoes. Meteorological disasters are caused by short-term atmospheric processes (i.e., from minutes to days) such as storms. Hydrological disasters are events caused by the deflections in the normal water cycle and/or overflow of water bodies due to the formation of winds such as floods and landslides. Climatological disasters are events caused by long-lived processes (in the spectrum of climate fluctuations during seasons to several decades) such as drought and wildfire. Biological disasters are events caused by living organisms' exposure to germs and toxic substances such as insect infestation [4–6]. This chapter focuses on geophysical, hydrological, meteorological, and climatological disasters.

In 2019, floods were the most common disaster type, with 194 floods recorded throughout the year. Ninety storms and 32 earthquakes, 25 landslides were recorded, while 21 extreme temperatures were recorded in 2019. Sixteen droughts, 14 wildfires, and 4 active volcanoes were recorded during 2019, as shown in Fig. 3. Floods were less deadly than earthquakes in terms of the number of lives lost due to the event.

Fig. 2 The main types of natural disasters

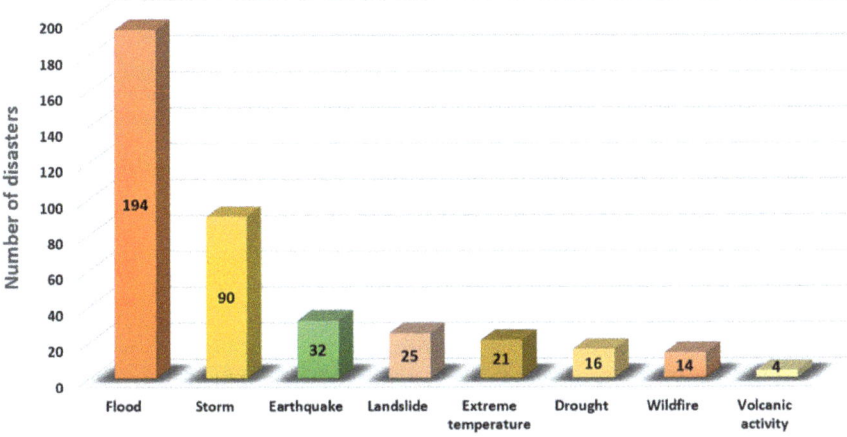

Fig. 3 Several natural disasters worldwide in 2019, by type. Data source [7]

However, the nature of catastrophic floods has changed in recent years. Flash floods are becoming more frequent, as are acute river and coastal floods. Urbanization has also greatly increased flood flow. Combined, these factors have raised the average death toll from floods in many parts of the world [6]. The death toll from the 2019 floods exceeded the earthquake's death toll, as the death toll from floods reached 5,100, while the death toll from earthquakes reached 259, as shown in Table 1. In 2018, the death toll from earthquakes was more significant than the death toll from floods, as 4,321 earthquake deaths were recorded, while 2,869 deaths were recorded due to floods. In 2019, the number of deaths caused by natural disasters reached 9,200, while in 2018, it was slightly lower, with 9,126 deaths recorded.

Disaster type	Number of deaths in 2018	Number of deaths in 2019
Flood	2869	5100
Earthquake	4321	259
Landslide	275	719
Extreme temperature	536	2908
Drought	0	77
Wildfire	247	116
Volcanic activity	878	21
Total	**9126**	**9200**

Table 1 Global deaths from natural disasters in 2018 and 2019. Data source [2]

2.2 Deep Learning Architectures

DL is a kind of machine learning method based on artificial neural networks. It has been proposed to bring machine learning closer to one of its main goals: artificial intelligence. DL has benefited from the increasing amounts of data and the continuous increase in available computer power. It has significantly improved artificial intelligence tasks such as image classification, machine translation, and many other tasks [8]. DL architectures' deep nature and their ability to extract critical features without the need for feature engineering have helped solve many of the most complex AI tasks [9, 10]. The use of DL architectures or other machine learning methods is essential to get some tasks done. For example, tasks in which humans cannot express their experiences, such as a speech recognition task or problem solution, change over time, such as tracking and weather forecasting tasks. Besides, tasks where the problem's magnitude is greater than humans' limited logical capabilities, such as the sentiment analysis task [11].

A large number of DL architectures can be used for various tasks such as the Deep Neural Network (DNN), which is considered the simplest deep learning architectures. Additionally, Convolutional Neural Networks (CNNs) are commonly used for computer vision tasks such as image classification and image localization [12]. Recurrent Neural Networks (RNN) including the two common types, Long Short Term Memory (LSTM), and Gated Recurrent Units (GRU) that are used to process sequential data. Generative Adversarial Networks (GAN) is used to generate new data samples [11]. Graph neural networks (GNNs) are relatively recent DL architectures that operate on the graph field [13].

3 Natural Disasters Concurrent with the COVID-19 Pandemic

As the world struggles to slow the spread of COVID-19, natural disasters have shown that they still loom large during the pandemic and expose many countries to additional risks from natural disasters. Usually, when a natural disaster strikes, mass evacuation is the only option to protect people from deadly storms, earthquakes and floods. However, under conditions of the COVID-19 pandemic, evacuating and sheltering millions of people is a major challenge and fertile environment for spreading this deadly virus. In this section, a review of some of the natural disasters that occurred in conjunction with the COVID-19 pandemic and their negative effects on the increase in infection with the COVID-19 are presented.

The Croatia earthquake is one of the first natural disasters to occur during the COVID-19 pandemic. The 5.4-magnitude earthquake that struck the northern suburbs of Zagreb, the capital of Croatia, on March 22, 2020, disrupted the lockdown and thus increased the number of COVID-19 cases [14]. Figure 4 shows a clear increase

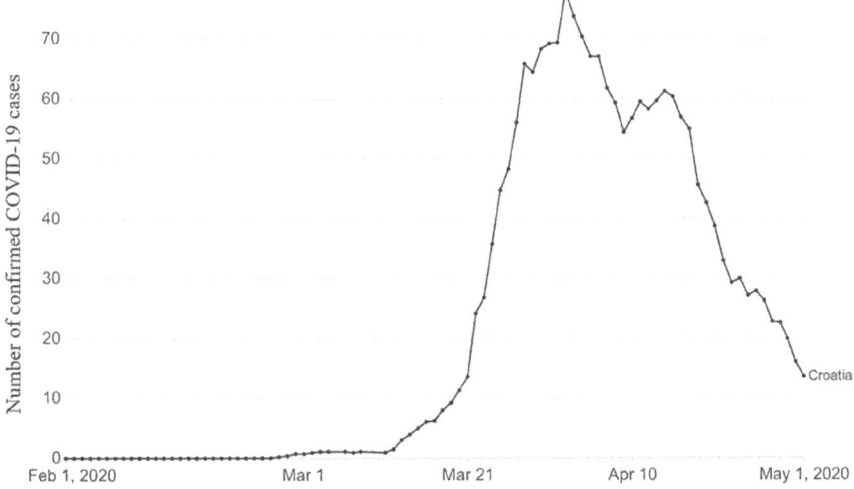

Fig. 4 Daily confirmed cases of COVID-19 in Croatia between January 1, 2020, and May 1, 2020. Data source [2]

in the number of people infected with COVID-19 after March 22, 2020, that is, in the aftermath of the Zagreb earthquake.

Among the most powerful natural disasters that occurred during the COVID-19 pandemic is the Super Cyclonic Storm Amphan. The Super Cyclonic Storm Amphan hit eastern India and Bangladesh's coasts in the third week of May 2020. The Super Cyclonic Storm Amphan left much damage, and more than 2 million people were evacuated from their homes in Bangladesh [15]. As a result of this necessary evacuation, the number of COVID-19 infections in Bangladesh has increased dramatically after mid-May 2020. As shown in Fig. 5, the daily confirmed cases of COVID-19 in Bangladesh on May 1, 2020, two weeks before the Super Cyclonic Storm Amphan formed, was about 497 cases. However, the number of confirmed COVID-19 cases in Bangladesh on May 30, 2020, about a week after the Super Cyclonic Storm Amphan ended, was around 1,805 cases. The number of confirmed cases of COVID-19 in Bangladesh on June 21, 2020, about a month after the Super Cyclonic Storm Amphan ended, was about 3,485 cases [2].

In conjunction with the Coronavirus crisis, the Kyushu floods occurred when record torrential rains fell in the Kumamoto and Kagoshima prefectures of Kyushu Island, southern Japan, on July 4, 2020. Due to the devastating floods and landslides that swept parts of Kyushu on July 4, 2020, 53 people have been confirmed killed as of July 7, 2020. Authorities ordered hundreds of thousands of people to evacuate their homes in Kumamoto and Kagoshima prefectures and directed them to several shelters in the area [16, 17]. Despite the authorities' efforts to maintain social distancing in the shelters to prevent the spread of the COVID-19 by reducing each shelter's carrying capacity while separating families with cardboard walls. Japan has witnessed a remarkable increase in the number of coronavirus infections after these floods and

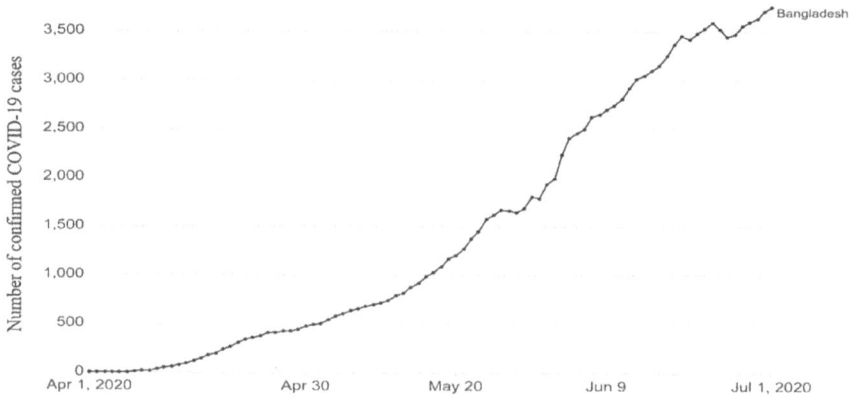

Fig. 5 Daily confirmed cases of COVID-19 in Bangladesh between April 1, 2020, and July 1, 2020. Data source [2]

landslides. As shown in Fig. 6, the daily confirmed cases of COVID-19 in Japan on July 1, 2020, seven days before the floods, were about 99 cases. However, the number of confirmed cases of COVID-19 in Japan on July 21, 2020, 14 days after the floods occurred, was about 502 cases. The number of confirmed cases of COVID-19 in Japan on August 7, 2020, a month after the floods occurred, about 1,434 cases [2].

Natural disasters are not limited to those mentioned above, but the year 2020 witnessed many natural disasters other than earthquakes and floods such as wildfires and active volcanoes, but the earthquakes and floods were the deadliest. Some of the earthquakes and floods that occurred in 2020 were a double disaster with the COVID-19 pandemic. As a result, the authorities in some countries were forced to

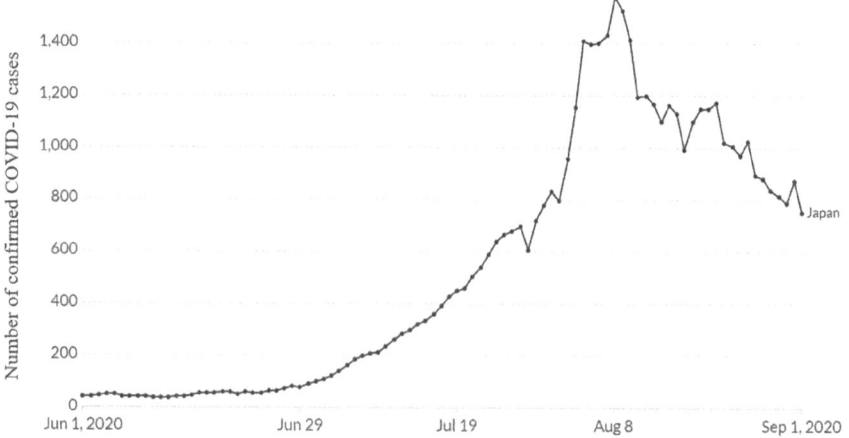

Fig. 6 Daily confirmed cases of COVID-19 in Japan between June 1, 2020, and September 1, 2020. Data source [2]

evacuate citizens and place them in evacuation shelters, which led to a significant increase in COVID-19 infection cases and the heavy loss of human life.

4 A Review of Deep Learning Applications in Natural Disaster Prediction

In light of the COVID-19 outbreak, early prediction of the occurrence of a natural disaster, especially earthquakes and floods, is of great importance to avoid an increase in the incidence of COVID-19. In addition to giving the authorities sufficient time to take action to reduce losses resulting from these disasters. Because floods and earthquakes were the most frequent and deadly occurrences in conjunction with the COVID-19 pandemic. A review of current deep learning techniques to predict the time, location, and magnitude of floods and earthquakes is conducted in this section.

4.1 Search Strategy

A literature search was conducted utilizing the Mendeley Scientific Database. For the group of studies focusing on flood forecasting using deep learning techniques, the search term "Flood Prediction by Deep Learning" was used on November 19, 2020. With this search term, 1038 articles were retrieved. To narrow the scope of the search, the document type was limited to "Article", and the year of publication was limited to "2019" and "2020". After this limitation, 60 articles were retrieved for analysis. To ensure the relevance and diversity of articles for flood prediction and deep learning, 60 articles were manually scanned by title, keywords, and abstract. If this could not be determined, the text was completely scanned. Finally, eight diverse and related articles were found and selected for in-depth analysis.

For the group of studies focusing on earthquake prediction using deep learning techniques, the search term "earthquake prediction through deep learning" was utilized on November 21, 2020. Forty-seven papers were obtained using this search term. To restrict the breadth of the search, the document type was restricted to "Article", and the year of publication was limited to "2019" and "2020". 23 papers were collected for review after this ban. To ensure the relevance and diversity of the articles for earthquake prediction and deep learning, 23 articles were manually scanned by title, keywords, and abstract. If this could not be determined, the full text was completely checked. Finally, five diverse and related articles were selected for in-depth review.

4.2 Deep Learning Applications for Early Flood Prediction

In this section, eight studies related to deep learning in the early prediction of floods are reviewed, as shown in Table 2. These studies were selected according to the methodology mentioned in Sect. 4.1.

In Dong et al. [18], a model called FastGRNN-FCN is proposed for the early prediction of flood sites based on sensor data on real-time precipitation and water levels. The FastGRNN-FCN model is based on two architectures of deep learning, the RNN and CNN. FastGRNN-FCN model can provide predictive flood warning and situational awareness at different time steps and locations with high efficiency. To measure the FastGRNN-FCN model's performance, the authors used several metrics such as accuracy and F-score. The accuracy and F-score of the FastGRNN-FCN model on test samples were 97.8% and 0.792%, respectively.

In Kim and Han [19], a DNN architecture was used to predict urban flooding, and the data augmentation technique was used to overcome data shortages and improve DNN performance. Through the trained DNN, the total amount of overflow of the drainage basin in Samseong-dong, Seoul, which was affected by heavy floods during the years 2010-2011, was predicted. The proposed methodology's performance was measured using the Mean Absolute Error (MAE), and this methodology

Table 2 Proposed approaches in literature studies for early flood prediction

The approach used	Objective
FastGRNN-FCN [18]	This model was proposed to provide early warning of potential flood sites in the next few hours based on sensor data on real-time precipitation and water levels
DNN with data augmentation [19]	This methodology was introduced to overcome the lack of data about the actual rainfall causing the floods and to create an ideal database using the data augmentation technique. The DNN model was also used to accurately predict the total cumulative surplus for the study area
GBDT [20]	The GBDT algorithm was used to provide a model that could predict the depth of flooded urban areas
DLNN [22]	This hybridization between deep learning and geographic information system is provided in order to be a promising tool to help mitigate flash floods
LSTM [23]	This model was used to predict precipitation, which is an important indicator of the likelihood of flooding
LSTM-seq 2seq [24]	This model was proposed to estimate the hourly rainfall runoff that often causes flooding
EMD-En-De-LSTM [25]	The main motivation for introducing this model is to provide continuous long-term forecasts and large flood forecasts
CNN [26]	This CNN based model was introduced for rapid prediction of river flood inundation

effectively forecasted urban floods. The authors stated that this methodology is useful in determining a flood risk area in an urban watershed.

In Wu et al. [20], a deep learning algorithm called a gradient Boosting decision tree (GBDT) has been used. The GBDT algorithm is a powerful deep learning technique used in classification and regression tasks and was proposed in Deng et al. [21]. The GBDT model is applied in conjunction with hydrological variables to predict the inundation and depths of flooded urban areas by learning and calculating the relationship between conditioning factors and the occurrence of floods. This model can be used to prevent and manage urban flooding. Several measures were used to measure the performance of the GBDT model, such as QR. QR denotes the ratio of the number of samples whose model simulation value does not exceed the permissible error (20%) for the total number of samples. The results showed that the GBDT model achieved great accuracy in predicting the depth of water accumulation for urban floods.

In Bui and Hoang [22], a classification model based on the deep learning neural network (DLNN) is proposed. It is possible to determine the probability of exposure of areas within the study area to flash flooding. Based on the collected Geographic Information System (GIS) database and the preliminary analysis based on Information Gain Ratio (InGR), the factors most influencing the occurrence of flash floods were identified. The performance of this proposed model was measured using several metrics such as accuracy. DLNN achieved 92.95% accuracy, outperforming other comparable approaches.

In Kang et al. [23], the LSTM, a type of RNN, was used. The LSTM model was used to forecast precipitation based on meteorological data from 2008 to 2018 in Jingdezhen. For best results, the used dataset variables that show a weak correlation between meteorological variables and the precipitation have been removed. The performance of this LSTM model was measured using several metrics such as RMSE. The results prove that the LSTM model can predict rainfall well, allow more flexible decisions to be made in Jingdezhen, and provides sufficient time to prepare strategies against potential flood damage.

In Xiang et al. [24], a model called LSTM-seq 2seq is proposed that relies on LSTM and Sequence to Sequencing, which are deep learning methods. The LSTM-seq 2seq model was provided to estimate the hourly rainfall runoff that often causes flooding. Because floods usually occur during periods of heavy rainfall that generate a lot of surface runoff. Rainfall-runoff modeling is very important in hydrological and environmental studies, so the LSTM-seq 2seq model and other similar models are very effective for flood forecasting and analysis. The LSTM-seq 2seq model was evaluated by comparison with other similar models, and the results proved the superiority of the LSTM-seq 2seq model over others. The results also indicated that the LSTM-seq 2seq has sufficient predictive power and can improve forecast accuracy in short-term flood forecasting applications.

In Liu et al. [25], a model called EMD-En-De-LSTM that based on the Empirical Mode Decomposition (EMD) algorithm and Encoder-Decoder Long Short-Term Memory (En-De-LSTM) architecture (En-De-LSTM) has been proposed. This hybrid model between EMD and En-De-LSTM helped make long-term predictions more efficiently. This model was proposed to predict the longtime (10 years) monthly streamflow data from Hankou Hydrological Station on the Yangtze River as a case study. The EMD-En-De-LSTM model's performance was evaluated using multiple criteria such as the Root Mean Square Error (RMSE). The results demonstrated the reliability of this model in catastrophic flood years and long and continuous prediction.

In Kabir et al. [26], an approach based on CNN is proposed. The proposed CNN method offers great potential for real-time flood modeling/forecasting. To ensure the proposed CNN's effectiveness, a comparison was made with machine learning methods such as the support vector regression method. The comparison result confirmed that the proposed method provides a promising tool for immediate flood prediction.

4.3 Deep Learning Applications for Early Earthquake Prediction

In this section, five studies related to deep learning applications in early earthquake prediction are reviewed, as shown in Table 3. These studies were selected according to the methodology mentioned in Sect. 4.1.

In Wang et al. [27], an LSTM network with two-dimensional input is constructed, which can detect spatio-temporal correlations between historical earthquake data and make predictions about earthquakes in a large area of interest. To measure the performance of the constructed LSTM network, the authors used several metrics such as accuracy. The results obtained showed that the proposed network could predict earthquakes with high accuracy.

Table 3 Proposed approaches in literature studies for early earthquake prediction

The approach used	Objective
LSTM [27]	This LSTM-based model was introduced to learn the spatial and temporal relationship among earthquakes in different locations and to take advantage of this relationship in predicting earthquakes better than before
CNN [28]	A CNN-based model has been proposed for predicting earthquake occurrence with high accuracy and addressing limitations of current methods
P-DNN [29]	This model was introduced to enhance regional damage assessments before and after earthquakes and to facilitate pre-earthquake risk mitigation efforts
GNN [30]	This approach has been proposed to improve the performance of machine learning methods in the seismic source characterization task (particularly in estimating location and magnitude)
LSTM-CNN [31]	This joint model has been proposed to predict the intensity function and the probability of a seismic event of a given magnitude occurring in a given area

In Jena et al. [28], a model based on CNN was proposed to estimate earthquakes' probability based on nine important indicators as inputs to train the model. Based on the inputs, the model determines whether or not an earthquake will occur. Several metrics were used to measure the model's quality, such as accuracy, precision, and recall. The results showed the accuracy of the CNN-based model in estimating an earthquake's probability, outperforming other conventional methods.

In Kim et al. [29], In this paper, a probabilistic deep neural network (P-DNN) model is proposed to overcome the challenges in assessing regional losses before and after the earthquake. These challenges are (1) mitigating pre-earthquake risks to reduce potential social and economic losses, (2) post-earthquake decision-making to reduce losses and enhance the potential for urban community recovery. The model's effectiveness and applicability have been comprehensively verified with real examples. The model's effectiveness and ability to meet the challenges above have been confirmed.

To define the location and magnitude of earthquakes, seismologists utilize seismic stations' geographical locations that register the earth's shaking in a data analysis workflow. However, most machine learning methods utilized in seismology do not take into account this spatial information. Thus there is an untapped potential for improving the performance of machine learning models. Therefore, in van den Ende and Ampuero [30], a method based on GNN has been proposed, which considers the geographical locations of seismic stations. The results show that the model achieves high prediction accuracy and opens up new applications in the automation of seismic tasks and earthquake early warning systems.

In Nicolis et al. [31], two deep learning architectures, LSTM and CNN, were used to predict the rate of earthquakes in Chile. This was done by relying on three units: a pre-processing unit, a spatial and time estimation unit, and a prediction unit. In the second unit, LSTM and CNN are used to predict earthquake intensity and location. The final unit integrates the information provided by these DL constructs to predict future values of the maximum rate and location of earthquakes. Several measures, such as accuracy, were used to verify the efficiency of the model. The results proved the effectiveness of the proposed model in predicting the rate of earthquakes with high accuracy.

5 Recommendations

Undoubtedly, government organizations, disaster management personnel, and local communities will face a difficult trade-off between saving people from the risks of natural disasters and reducing the spread of COVID-19. Therefore, it can be argued that the time has come to rely on seasonal prediction models based on machine learning. As it is possible that these machine learning-based models can contribute significantly to predicting the occurrence of natural disasters days or weeks before their occurrence, which allows intensifying contingency planning procedures. Moreover, local governments should develop an evacuation plan to deal with devastating natural disasters such as the super cyclonic storm that occurred in Bangladesh, which was mentioned before. Local governments should create evacuation centers that accommodate millions of people and allow for social distancing. Governments should be keen to allocate field hospitals to treat the wounded and injured due to natural disasters, far from those designated to deal with those infected with the COVID-19. Local governments must also ensure the safety of their disaster management personnel. Governments should also require staff and volunteers to implement social distancing while managing natural disasters to avoid putting them at risk of contracting COVID-19 infection. It is also necessary to involve scientists in health, disaster, and crisis management, in developing plans and decision-making processes in light of these exceptional critical circumstances.

6 Conclusion and Future Work

The coincidence of natural disasters with pandemics exacerbates the two crises, as the effects of the natural disaster will be worse than they might otherwise be, and the additional outbreak of the pandemic. For this reason, this chapter has introduced a statistical analysis approach to the danger of compounding influences originating from natural disasters during the COVID-19 pandemic. Also, the literature review approach for the DL applications in natural disaster prediction has been presented. Through these two approaches, we can present a set of recommendations that may assist the governments in developing strategies for risk reduction of natural disasters and reducing the spread of COVID-19 using AI applications. In future work, to expand the literature search base, the search will be conducted utilizing more scientific databases such as Scopus and Google Scholar. Also, more types of natural disasters, more AI techniques, and applications will be extensively studied.

References

1. World health organization. https://www.who.int/emergencies/diseases/novel-coronavirus-2019 (2020). Accessed 6 Dec 2020
2. Roser, M., Ritchie, H., Ortiz-Ospina, E., Hasell, J.: Our world in data. Statistics and research, Coronavirus (COVID-19) cases. https://ourworldindata.org/covid-cases (2020). Accessed 21 May 2020
3. Goswami, S., Chakraborty, S., Ghosh, S., Chakrabarti, A., Chakraborty, B.: A review on application of data mining techniques to combat natural disasters. Ain Shams Eng. J. 9(3), 365–378 (2018)
4. Guha-Sapir, D., Vos, F., Below, R., Ponserre, S.: Annual disaster statistical review 2011: the numbers and trends. Centre for Research on the Epidemiology of Disasters (CRED) (2012)
5. Smith, K.: Environmental hazards: assessing risk and reducing disaster. Routledge January 16 (2013)
6. UNISDR, CRED.: The human cost of natural disasters: a global perspective (2015)

7. Statista.com.: Number of natural disasters worldwide by type 2019, Statista Inc. Retrieved https://www.statista.com/statistics/269653/natural-disasters-on-the-continents-by-nature-of-the-disaster/ (2020)
8. LeCun, Y., Bengio, Y., Hinton, G.: Deep Learning. Nature **521**(7553), 436–444 (2015)
9. Erickson, B.J., Korfiatis, P., Kline, T.L., Akkus, Z., Philbrick, K., Weston, A.D.: Deep learning in radiology: does one size fit all? J. Am. College Radiol. **15**(3), 521–526 (2018)
10. Darwish, A., Ezzat, D., Hassanien, A.E.: An optimized model based on convolutional neural networks and orthogonal learning particle swarm optimization algorithm for plant diseases diagnosis. Swarm Evol. Comput. **1**(52), (2020)
11. Alom, M.Z., Taha, T.M., Yakopcic, C., Westberg, S., Sidike, P., Nasrin, M.S., Asari, V.K.: A state-of-the-art survey on deep learning theory and architectures. Electronics **8**(3), 292 (2019). https://doi.org/10.3390/electronics8030292
12. Ezzat, D., Hassanien, A.E., Ella, H.A.: An optimized deep learning architecture for the diagnosis of COVID-19 disease based on gravitational search optimization. Appl. Soft Comput. **22**, (2020)
13. Zhou, J., Cui, G., Zhang, Z., Yang, C., Liu, Z., Wang, L., Li, C., Sun, M.: Graph neural networks: a review of methods and applications. arXiv:1812.08434 (2018). Accessed 20 Dec 2018
14. Quigley, M.C., Attanayake, J., King, A., Prideaux, F.: A multi-hazards earth science perspective on the COVID-19 pandemic: the potential for concurrent and cascading crises. Environ. Syst. Dec. (2020). https://doi.org/10.1007/s10669-020-09772-1
15. Baidya, D.K., Maitra, S., Bhattacharjee, S.: Facing post-cyclone disaster in times of COVID-19 Pandemic in India: possible testing strategy to reduce further spread of disease. Asia Pacific J. Public Health. **32**(6–7), 376 (2020)
16. The Japan Times.: At least 53 dead as torrential rains and floodwaters hit Kyushu. The Japan Times, July 7. https://www.japantimes.co.jp/news/2020/07/07/national/rain-floods-kyu shu/ (2020)
17. NASA Earth Science Disasters Program.: NASA measures flooding rainfall from plum rains in Japan. NASA Earth Science Disasters Program, July 7. https://gpm.nasa.gov/applications/ weather/imerg-measures-flooding-rainfall-plum-rains-japan (2020)
18. Dong, S., Yu, T., Farahmand, H., Mostafavi, A.: A hybrid deep learning model for predictive flood warning and situation awareness using channel network sensors data. Comput. Aided Civil Infrastr. Eng. (2020). Accessed 15 June 2020
19. Kim, H.I., Han, K.Y.: Urban flood prediction using deep neural network with data augmentation. Water **12**(3), 899 (2020). https://doi.org/10.3390/w12030899
20. Wu, Z., Zhou, Y., Wang, H., Jiang, Z.: Depth prediction of urban flood under different rainfall return periods based on deep learning and data warehouse. Sci. Total Environ. 137077 (2020). https://doi.org/10.1016/j.scitotenv.2020.137077
21. Deng, S., Wang, C., Wang, M., Sun, Z.: A gradient boosting decision tree approach for insider trading identification: an empirical model evaluation of China stock market. Appl. Soft Comput. **83**, 105652 (2019)
22. Bui, D.T., Hoang, N-D., Martínez-Álvarez, F., Ngo, P-T.T., Hoa, P.V., Pham, T.D., Costache, R.: A novel deep learning neural network approach for predicting flash flood susceptibility: A case study at a high frequency tropical storm area. Sci. Total Environ. 134413 (2019)
23. Kang, J., Wang, H., Yuan, F., Wang, Z., Huang, J., Qiu, T.: Prediction of Precipitation Based on Recurrent Neural Networks in Jingdezhen, Jiangxi Province. China. Atmosp. **11**(3), 246 (2020). https://doi.org/10.3390/atmos11030246
24. Xiang, Z., Yan, J., Demir, I.: A rainfall runoff model with LSTM based sequence to sequence learning. Water Res. Res. **56**(1), e2019WR025326 (2020)
25. Liu, D., Jiang, W., Mu, L., Wang, S.: Streamflow prediction using deep learning neural network: case study of yangtze river. IEEE Access **8**, 90069–90086 (2020). https://doi.org/10.1109/acc ess.2020.2993874
26. Kabir, S., Patidar, S., Xia, X., Liang, Q., Neal, J., Pender, G.: A deep convolutional neural network model for rapid prediction of fluvial flood inundation. J. Hydrol. 125481 (2020)
27. Wang, Q., Guo, Y., Yu, L., Li, P.: Earthquake prediction based on spatio-temporal data mining: an LSTM network approach. IEEE Trans. Emerg. Topics Comput. (2017). Accessed 27 April 2017

28. Jena, R., Pradhan, B., Al-Amri, A., Lee, C.W., Park, H.: Earthquake probability assessment for the indian subcontinent using deep learning. Sensors **20**(16), 4369 (2020). https://doi.org/10.3390/s20164369
29. Kim, T., Song, J., Kwon, O.S.: Pre-and post-earthquake regional loss assessment using deep learning. Earthquake Eng. Struct. Dynam. **49**(7), 657–678 (2020)
30. van den Ende, M.P., Ampuero, J.P.: Automated seismic source characterization using deep graph neural networks. Geophys. Res. Lett. **47**(17), e2020GL088690 (2020). Accessed 16 Sep 2020
31. Nicolis, O., Plaza, F., Salas, R.: Prediction of intensity and location of seismic events using deep learning. Spatial Stat. **14**, (2020)

Sustainable Climate Change Policies Driven by Global CO_2 Reduction During COVID-19

Haytham H. Elmousalami

Abstract In Wuhan, the Coronavirus has exploded into an international pandemic. China quarantined millions of people and shut down its economy. Italian leaders grounded flights, suspended mortgage payments, and cleared streets with mandatory lockdowns. US President Trump unilaterally banned all travel from European countries. This international confinement limits human activities in all aspects of life such as power, Industry, transportation, aviation, and work. As a result, global CO_2 emissions have been dropped. Accordingly, the COVID-19 response and policies have a positive impact on the environment and climate change. Therefore, this study analyses the change in CO_2 emissions during the COVID-19 pandemic and presents a set of Sustainable Climate Change Policies (SCCPs) based on COVID-19 Practices to maintain the United Nations (UN) Sustainable Development Goals (SDG). The more effective path is moving towards more energy efficiency and less carbon per unit of economic activity. Furthermore, the response to the COVID-19 demonstrates that planned economic slowdowns are not only possible but necessary to cut emissions drastically.

Keywords COVID-19 · Climate change · Sustainable climate change policies (SCCPs) · CO_2 reduction · Zero carbon future · Artificial intelligence · Lessons learned

1 Introduction

The 2008 economic recession decreases the greenhouse gas emissions when the economy slows down in the past. Similarly, the collapse of the Soviet Union and the Eastern European block, there is a significant reduction in greenhouse gas emissions

H. H. Elmousalami (✉)
Project Engineer, Project management professional PMP at General Petroleum Company (GPC), Cairo, Egypt
URL: http://www.egyptscience.net

Researcher at Scientific Research Group in Egypt (SRGE), CAIRO, Egypt

© The Author(s), under exclusive license to Springer Nature Switzerland AG 2021
A. E. Hassanien et al. (eds.), *The Global Environmental Effects During and Beyond COVID-19*, Studies in Systems, Decision and Control 369,
https://doi.org/10.1007/978-3-030-72933-2_8

over five-year periods or even longer associated with this really kind of cataclysmic reduction in economic activity. Earth has a big fire season in the western US and a big hurricane in the East. Humans worry about climate shocks whether and the simultaneous crop failures in multiple regions or a year. However, these climate interruptions are normal results of current climate change [1]. Therefore, there is a need for a sustainable solution against climate change and its circumstances. In 2018, the UN's Intergovernmental Panel on Climate Change reported that the United States (USA) and other prolific emitters need to transition away from fossil fuels and cut carbon emissions to Net Zero by 2050 to avoid a level of global warming that can be seriously harmful to the environment and human health [2, 3].

Recently, The COVID-19 pandemic causes a global crisis that has wide-ranging impacts across all stratum of society. Therefore, Humans have to rapidly adapt and change their behaviors in unexpected and unprecedented ways [4]. The impacts of the COVID-19 pandemic will be lasted for generations, both emotionally and financially. Therefore, The COVID-19 pandemic has a significant effect on the UN Sustainable Development Goals (SDG) (i.e., Environment, Energy, and Social Equity) [5]. The US Senate passed a two trillion-dollar stimulus bill called the coronavirus aid relief and Economic Security Act. It is the largest economic relief package in modern American history. It includes direct payments to Americans, an expansion of unemployment, and aid to large and small businesses [6]. Shutting down of the transportation sector, Industry, and the broader economy indicates that many critical pollutants are dropping in their concentrations to 30%. The air quality is rapidly improving in those places that are responding to market reductions. For example, California is going under many of these restrictions, air quality improving by 30% [7]. This economic reduction of 10–30% in northeast China and the industrial part of northern Italy improves the global air quality and environmental conditions.

Climate changes making naturally occurring phenomena like storms and fires. Ecological destruction is making the same thing for diseases. Humans might have infected several diseases like COVID-19 naturally because humans break down these ecosystems and mess with animals in the wrong ways. Therefore, humans are exposed to more and more of these kinds of biological storms. They are entering the human population to a biological crisis such as recent COVID 19 pandemics. While human activity does not cause diseases like the Coronavirus. It is certainly linked to their emergence of other diseases like Zika and malaria. These diseases are also tied to human activities as the climate gets warmer. These diseases are carried by mosquitoes [8]. Tropical climates have more diseases than cooler climates. Human activity can exacerbate a disease once it is taken place. For example, SARS was a respiratory disease that humans saw almost 20 years ago. If SARS infects a human, it is expected to die twice as likely from SARS if the patient lived in a city that was polluted by the fossil fuels before the Coronavirus.

In the recent past, viruses are naturally occurring in ecosystems. However, human activity increases the chances that a disease will jump from an animal population to humans. The virus originated in the animal population, where scientists believe that it was a bat. The genome of the Coronavirus and bats is about 96% identical to humans. It was then transmitted through an intermediary animal, perhaps a Pangolin, sold at a China wildlife market [9, 10]. The virus jumped to humans through a process known as zoonosis. The novel Coronavirus first emerged in humans in late 2019. A study published on Apr 1 looked at the contact between humans and primates in western Uganda. The authors looked at the satellite data. Researchers found that forests' destruction increases human-animal contact where the virus chances would be transmitted from animals to humans [11]. Therefore, humans contacting wildlife increases the risk of those diseases jumping from animal to human populations. Thus, Humans must respect the environmental balance and maintain the natural ecosystem to avoid biological storms and the global warming.

2 CO_2 Emissions Reduction During the COVID-19

A greenhouse gas (GHG) such as carbon dioxide (CO_2), ozone (O_3), methane (CH_4), nitrous oxide (N_2O), and water vapor (H_2O) absorbs and reflects radiant energy to Earth's atmosphere [12]. As a result, the greenhouse is caused where the average temperature of Earth's surface is increased from to -18 °C (0 °F) to 15 °C (59 °F). The main reason for this temperature increase is the human activities during the industrial revolution (around 1750) where the atmospheric concentration of CO_2 is increased by 45% (from 280 ppm in 1750 to 415 ppm in 2019). Energy generated by the combustion of fossil fuels such as coal, oil, and natural gas is the main cause of CO_2 emissions in our plant. Thus, catastrophic consequences of climate change will face human humanity unless the GHG emissions are reduced during the few coming years (IPCC 2018). The Coronavirus is the first and foremost human tragedy millions of people have been infected, and a hundred of thousands of people have died worldwide. Economies are shuttering, and people are suffering. However, there are positive environmental effects. Experts estimate that carbon emissions in China fell by 25% between February and March. In March, NASA satellite data found a 30% drop in air pollution over the Northeast United States, the lowest for any March since 2005 [13].

2.1 Data Collection

CO_2 emissions data are crucial to undertint the impact of global warming and variations in climate change. However, CO_2 emissions are not reported daily but in annual reports [14]. Moreover, satellite records of CO_2 emissions data have a large marginal error and several noisy data. Therefore, the recorded data have large uncertainty. There is a need for a global system to measure and monitor global CO_2 emissions in real-time with low uncertainty. Accordingly, a confinement index (CI) is an alternative approach to assess the CO_2 emissions at the country level due to the lack of real-time CO_2 emissions data [15]. This study compared the change in CO_2 emissions daily to assess the relative change against the pre-COVID conditions quantitatively.

The analysis is conducted for 69 countries representing 97% of global CO_2 emissions and 85% of the world population. The data provides a quantitative indication of CO_2 emission based on the activity changes in six economic sectors: power, Industry, surface transport, public, residential, and aviation. The change of CO_2 emissions is calculated based on the following Eq. (16):

$$\Delta CO_2^{c,s,d} = CO_2^c \times \mu S^c \times \Delta A^{s,d(CI,c)} \tag{1}$$

Where

- ΔCO_2: changes in emissions in $MtCO_2\ d^{-1}$
- c,s,d: for each country/state/province (c), sector (s) and day (d)
- CO_2^c: the daily average CO_2 emissions for the latest year
- (2019) updated based on the world countries Global Carbon Project (Friedling-stein et al. 2019).
- μS^c: The fraction of CO_2 emissions in each sector using the International Energy Agency (IEA) data.
- $\Delta A^{s,d(CI,c)}$: the activity level change for each sector compared with pre-COVID Levels.

2.2 Data Results and Analysis

Temporary reduction exists in daily global CO_2 emissions during the COVID-19 crises [15]. In April 2020, daily global CO_2 emissions were decreased by −17% (−11 to −25% for ± 1σ) compared with the mean 2019 levels because of confining populations to their homes and closing international borders. Figure 1 shows the change in country fossil CO_2 emissions (percent). The analysis is conducted for 69 countries (97% of global emissions). The 69 countries have been classified into five categories: China, Europe, India, the USA, and the rest of the world. On Dec 30, 2019, the emergence of COVID-19 began in Wuhan in China and rapidly spread worldwide. Chine's CO_2 emissions started to decline at the beginning of January 2020 and reached their peak with −6.0%, and then Chine's CO_2 emissions reached their normal percentages in May 2020. As a result, the reduction in CO_2 emissions in China is reflected by the activity reduction due to the COVID-19 spread (Fig. 2).

On Mar 11, 2020, the World Health Organization (WHO) declared that COVID-19 is a global pandemic [17]. Similarly, the COVID-19 reached Europe, where Europe's CO_2 emissions reduction peaked with −4.0% as a stacked percentage in mid of March 2020. The Europe's CO_2 emissions reduction is still existing due to the social distance precautions against COVID-19. In the USA, CO_2 emissions reduction reached −10% in April 2020 due to the COVID-19 precautions. Therefore, there is a strong relation between CO_2 emissions reduction, COVID-19 spread, and human activities.

The world currently emits approximately $100\ MtCO_2\ d^{-1}$. The CO_2 emissions are steadily increasing from approximately $20\ MtCO_2\ d^{-1}$ in 1950 to $100\ MtCO_2\ d^{-1}$ in

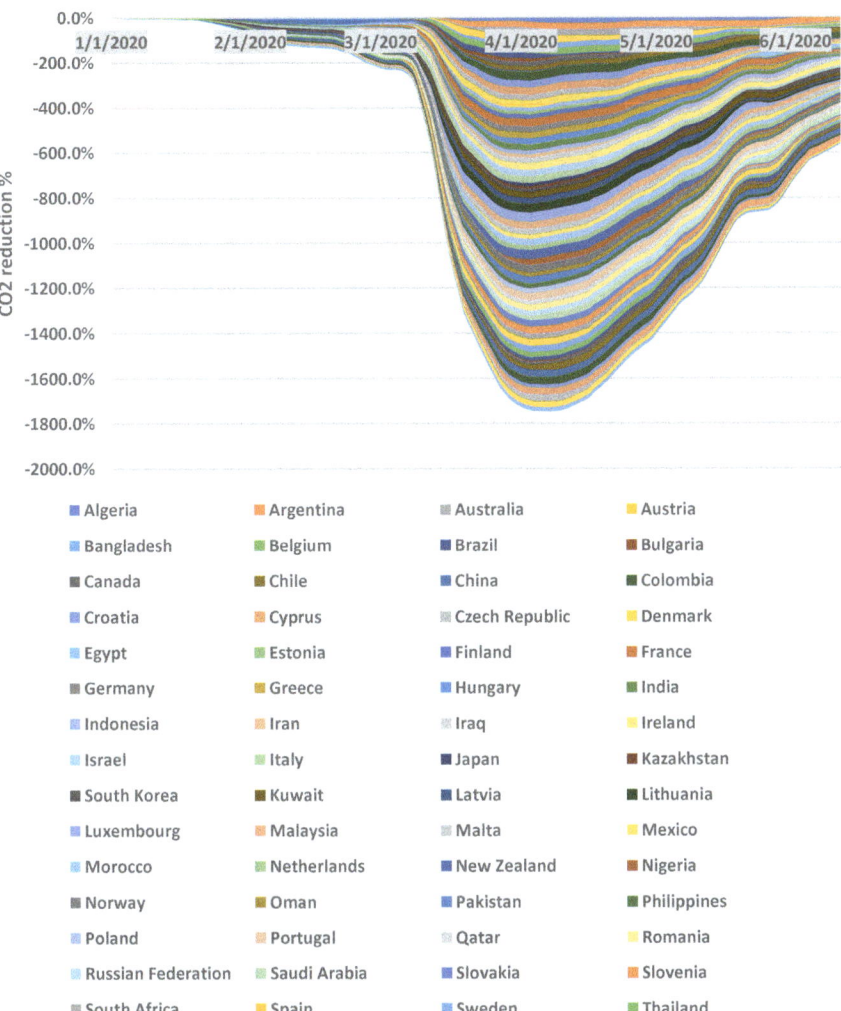

Fig. 1 Spectrum of stacked change in country fossil CO_2 emissions (percent)

the beginning of 2020. This CO_2 increase is the key reason for global climate change and global warming. As illustrated in Fig. 3, the annual mean daily CO_2 emissions in 1959–2020 (middle line) with the uncertainty of $\pm 5\%$ ($\pm 1\sigma$; low uncertainty and high uncertainty). CO_2 emissions are rising by about 1% per year between 1960 to 2020 before COVID-19, as shown in Fig. 3.

Historically, the changes in CO_2 emissions are driven by economic crises and energy factors. In 1973, there were oil crises where several industrial activities in the western world have shut down. Similarly, the global financial crisis was exploded in 2008, where the CO_2 emissions were decreased by -1.4% in 2009. However, in 2010,

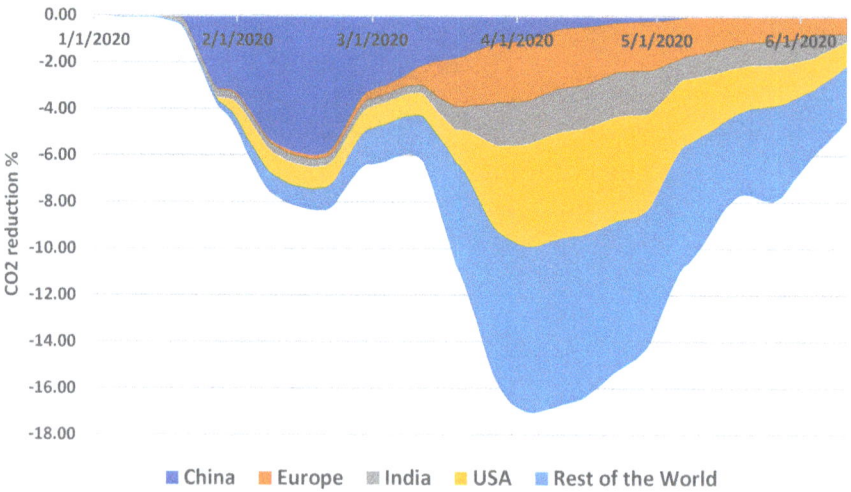

Fig. 2 Stacked change in global daily fossil CO_2 emissions attributed to each country (%)

this situation was followed by a growth in CO_2 emissions by $+5.1\%$ [18]. On Dec 30, 2019, the emergence of COVID-19 began in Wuhan in China and rapidly spread worldwide. On Mar 11, 2020, the World Health Organization (WHO) declared that COVID-19 is a global pandemic [17]. The pattern of human economic activities, world shipping, transportation, human economic activities, energy production, and demand has drastically changed. Consequently, the levels of CO_2 emission have also drastically changed from 100 to 80 $MtCO_2$ d^{-1} in Apr 4 2020 as showed in Fig. 3.

The data provides a quantitative indication of CO_2 emission based on the activity changes in six economic sectors: power, Industry, surface transport, public, residential and aviation. As illustrated in Fig. 4, the power and industry sectors represent 44.3% and 22.4% of global fossil CO_2 emissions, respectively. The surface transport and public buildings and commerce (public) sectors represent 20.6% and 4.2%, respectively (Liu et al. 2020). The residential and aviation sectors represent 5.6% and 2.8%, respectively. The key sectors for global fossil CO_2 emissions are power, Industry, and surface transport which they represent 87.3% of the CO_2 emissions based on Pareto analysis as displayed in Fig. 4.

CO_2 emissions change based on each sector's activity change pre-COVID-19 and during the COVID-19 pandemic. The power sectors' changes are driven by the electricity and power supply and demand in Europe, United States, and India. The changes in surface traffic and aviation sectors are driven by each country's historical traffic data [19]. The largest decrease in the aviation sector and surface transport by -75% and -50%, respectively. On the other hand, the industry and public sectors have lowered by -35% and -33%, respectively. The power sector has decreased by -15%. Conversely, the residential sector has increased by $+5\%$ due to staying at home attitude, as showed in Fig. 5.

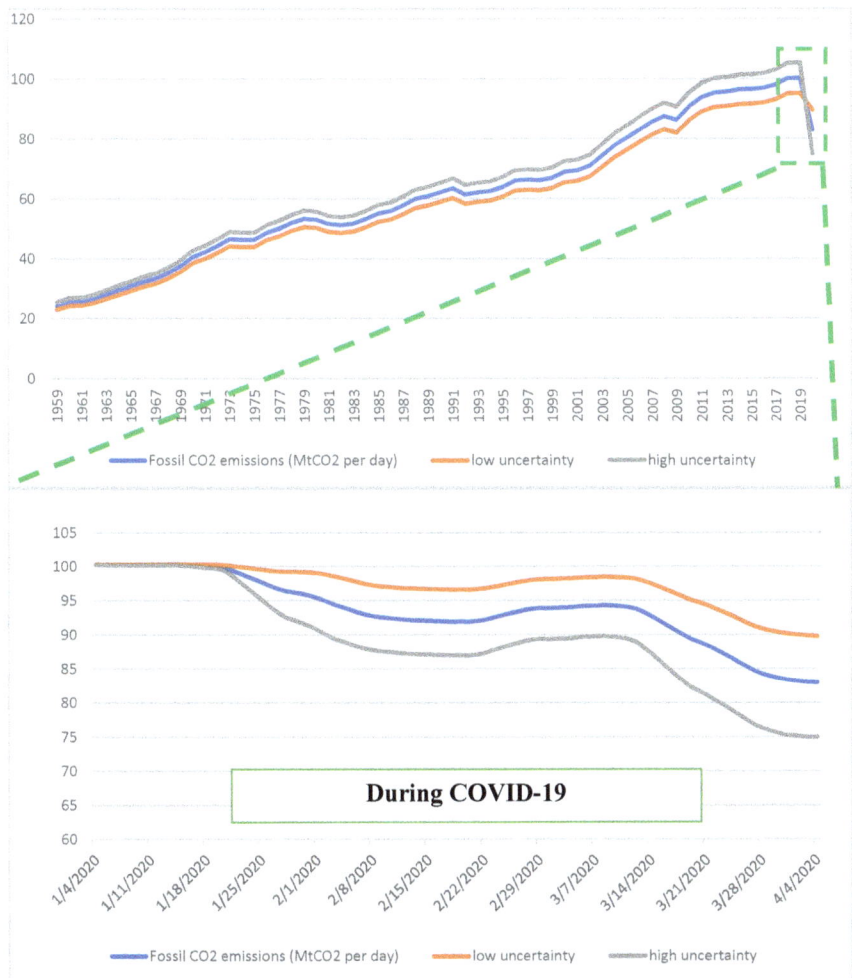

Fig. 3 Daily global fossil CO_2 emissions ($MtCO_2$ d^{-1})

China and United States (USA) are the highest CO_2 emitters in our plant. By focusing on transportation changes during the COVID-19 pandemic, international aviation has decreased by -75% in China during the peak of CO_2 emissions reduction. Similarly, global aviation has decreased by -20% in the USA. On the other hand, international shipping has decreased by -20% in China during the peak of CO_2 emissions reduction. Similarly, international shipping has approximately decreased by -75% in the USA. This pattern reflects the relation between CO_2 reduction, international shipping, and global aviation due to the COVID-19 pandemic (Fig. 6).

Due to the economic effects and recession impact of the COVID-19 crises, many countries take corrective actions against social distance and complete borders isolations. In China and Europe, transportation and aviation started to recover after a

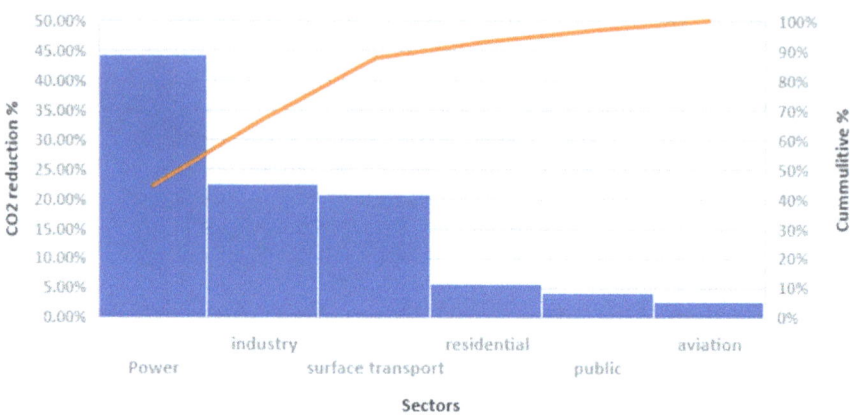

Fig. 4 Global fossil CO_2 emissions by sectors

semi-control of COVID-19 spread. Gradually, human activities began to heal after COVID-19 impact in several aspects of human activities. Consequently, the CO_2 emissions returned to their pre-COVID-19 rate at approximately 100 $MtCO_2\ d^{-1}$ in August 2020, as showed in Fig. 7. Therefore, the climate change challenge will be the key challenge for humanity after the COVID-19 confinement.

3 Lesson Learned from COVID-19 Crises for Climate Change Fighting

Climate change is making certain extreme weather events more likely, raising the risk of death either directly from that extreme event or indirectly through things that extreme event contributes to. There is a clear connection between emissions rates and the economy. The NASA satellite images show the steep drop in Nitrogen Dioxide air pollution during the sharp decline in economic activity due to China's rapid and aggressive response to the Coronavirus [20]. Carbon Brief asserts that China's coronavirus shutdown temporarily decreased the country's CO_2 emissions by a quarter, which Stanford Professor Marshall Burke predicts might have possibly reduced the number of premature deaths due to air pollution. Therefore, China's [21]. Overall mortality rate might have decreased in the two months during the height of the coronavirus shutdowns.

The point is not that pandemics are reasonable or necessary. It is instead that there is a significant, hidden toll of fossil fuel emissions. However, to prevent the millions of future deaths caused directly through burning fossil fuels or indirectly through the consequences of a hotter planet, the world needs to act quickly to create rapid and drastic structural change. The often-quoted Intergovernmental Panel on Climate Change report asserts that Humans have until 2030 to make sharp global emission cuts, which many argue is impossible. The Coronavirus definitively shows

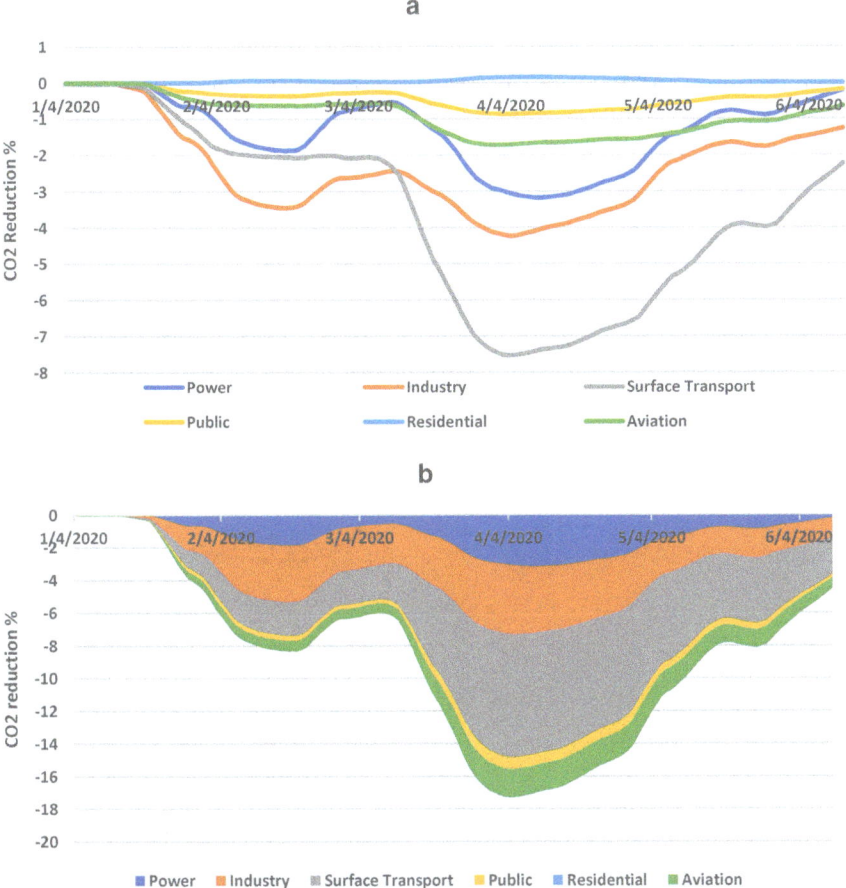

Fig. 5 **a** Percentage change and **b** Stacked change in global daily fossil CO_2 emissions attributed to each sector

that collective, large-scale, structural change is feasible in the face of a crisis. On the other hand, climate change is the biggest crisis of the current generation. As Amy Jaffe, director of the Council on Foreign Relations' Energy Security and Climate Change program, puts it, "Suppose you were a policymaker, and you were thinking about what you would do to lower emissions" [21].

Humans and governments have responded globally to the Coronavirus, where China, South Korea, Singapore, Italy, and the United States have quickly mounted large governmental responses to the Coronavirus. COVID-19 is a serious global and political crisis; however, the Coronavirus shows that rapid behavioral and structural change is possible in the face of the crisis. The response to Coronavirus (COVID-19) has been quick and broad. The response to climate has been slow and small. If humans and governments treated climate change like they have COVID-19, The

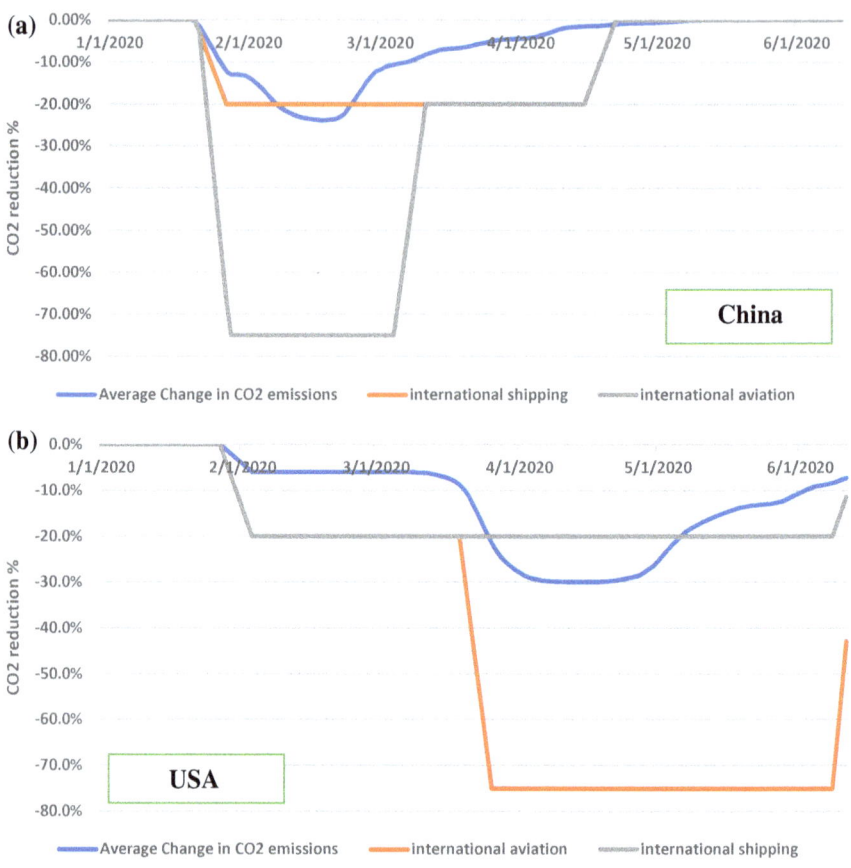

Fig. 6 **a** Change in China Province-level fossil CO_2 emissions (percent). **b** Change in US state-level fossil CO_2 emissions (percent)

way to zero-carbon practices would be the future. There are key significant shared challenges between climate change crises and COVID-19 as the following:

1. Delayed actions are high costly: delayed actions have catastrophic consequences, increase the mitigation, limit the possible mitigation and policy options.
2. Collaborate for better outcomes: international communities should collaborate to avoid global climate change based on international policy supported by United Nations (UN) Sustainable Development Goals (SDG).
3. Scientific contribution is the key for optimal solutions: more studies should be considered as a policy standard to fight climate change based on scientific facts using strict policies.
4. Citizens' behaviors and awareness: citizens and individuals are the units of any community. Therefore, intensive global and local media and educational

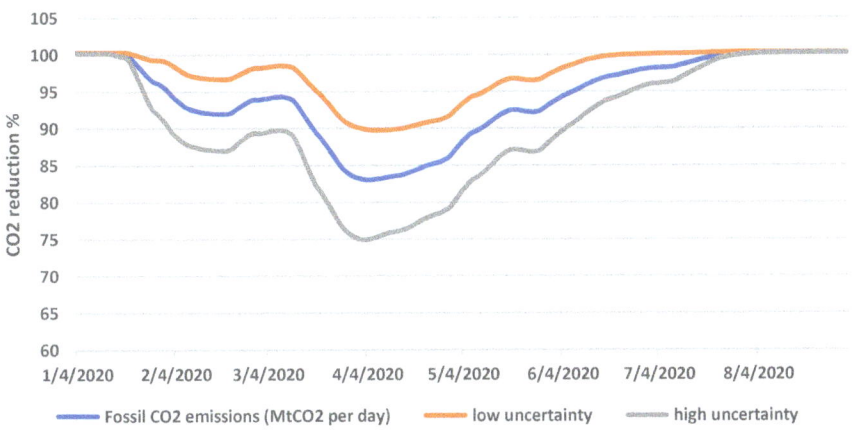

Fig. 7 CO_2 emissions during COVID-19 pandemic

awareness are required to enhance community support against climate change and CO_2 emissions.

4 Proposed Climate Change Policy Based on COVID-19 Crises

Facing climate change includes mitigation (eliminating GHG emissions) and adaptation (preparing for unavoidable consequences). Mitigation of GHG involves sustainable energy systems, zero energy building technology, and changing land use [22]. Adaptation necessities crises, disaster management, and extreme events understanding. Accordingly, public energy policies are required to complement and support energy use and generation in the building sector. The policy can refer to best practices, measures, laws, or standards executed by international organizations, standards committees, governments, or professional institutions.

The Coronavirus, countries such as Italy have almost stopped travel, many previously busy streets are recently free of cars. Workweeks are shortening, others are embracing the potential of video chat and messaging software instead of traveling long distances, and some companies have staggered work shifts to reduce traffic. Temporary bike lanes were set up all over New York City, and walking and biking were encouraged over other transportation options. The answer to climate change is not to quarantine everyone in their homes. That would be an absolute disaster. The Coronavirus response demonstrates that planned economic slowdowns are not only possible but necessary to cut emissions drastically. However, this type of fast structural change shows that without robust social safety nets like a jobs guarantee, extensive free public transit, or a strong low-carbon, low-cost public housing system, degrowth will harm millions of workers. Climate action propositions like the Green

New Deal (GND) need to incorporate this type of essential framework in their poli-cymaking because to combat climate change quickly; humans need a rapid structural transition [23].

The data analysis of CO_2 emission data illustrated the positive impact of COVID-19 on our plant healing, where daily global CO_2 emissions have been decreased by −17%. Moreover, emissions in individual countries have been decreased by −26% on average. This impact has been caused using the following policies:

1. Closure of selected national borders and restricted international travel.
2. Ban public gatherings and mandatory closure of schools, public buildings, universities, religious or cultural buildings.
3. Mandatory national 'lockdown', enforce social distancing.
4. Travelers self-quarantine from affected countries.
5. Isolating of the sick or symptomatic individuals.

The application of these COVID-19 emergency procedures and policies causes decreasing CO_2 emissions. However, by controlling of COVID-19 pandemic, the CO_2 emissions will increase again to its previous rate causing the climate change impact on Earth. The key objective is to limit the climate change to a 1.5 °C warming in the next decades [24]. The more effective path is moving towards more energy efficiency and less carbon per unit of economic activity [25]. Therefore, this study suggests Sustainable Climate Change Policies (SCCPs) that depend on the COVID-19 policies, sustainability policies, and artificial intelligence (AI) as showed in Fig. 8. The policies have divided into short-term policies and long-term policies as in Table 1.

COVID-19 practices induce short-term policies such as reducing transportation, planned economic slowdowns, electronic work, and education. Application of these COVID-19 practices as rigorized policics will reduce CO_2 emissions and work

Fig. 8 Sustainable climate change policies (SCCPs) using COVID-19 practices and artificial intelligence

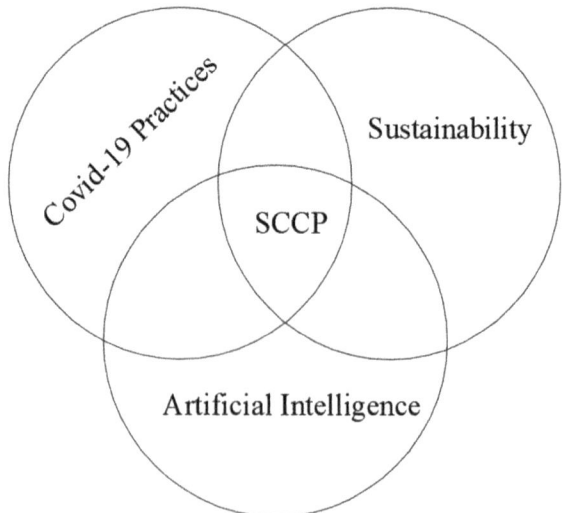

Table 1 The Short-term and long-term policies categorization

Short-Term Policies	Long-Term Policies
1. Reducing transportation activity (International shipping and international aviation)	1. 100% renewable energy production [36]
2. Reducing work hours outside homes activity	2. Zero Energy building [37]
3. Applying and generalizing electronic education	3. Smart green cities [38]
4. Applying and generalizing electronic home work	4. Electrical vehicles [39]
5. Applying carbon tax [40]	
6. Optimizing building energy	
7. Increasing renewable energy against fossil fuels. Boosting renewable energy based on AI and machine learning	
8. Planned economic slowdowns	

against climate change. Moreover, Generalizing and applying Carbon tax will boost citizens' and corporations' participation to minimize CO_2 in their daily lifestyle [40].

Yearly, electricity systems cause 25% of global greenhouse gas emissions [26]. Moreover, 40% of energy is consumed by buildings where 36% of the CO_2 emissions are generated by buildings [27, 28]. Therefore, Global society must rapidly take corrective decisions towards low-carbon electricity sources (such as solar, wind, hydropower, and geothermal energy) instead of carbon-emitting sources. Improving buildings' energy performance will effectively reduce the gas emissions, save energy and operational costs against the climate change issue [29, 30]. To solve the problem of climate change, energy consumption and CO_2 emissions. There is a worldwide trend to minimize CO_2 emissions by minimizing energy consumption based on a series of energy policies. For example, China is the first responsible for CO_2 emissions where China is the biggest emitter in our plant. The main reason for CO_2 emissions is that 59% of total energy consumption was coal consumption in 2018 [28]. Therefore, the Chinese government has applied several standards and policies based on green buildings, passive ultra-low energy green buildings, and nearly zero-energy buildings (NZEBs).

Machine learning and artificial intelligence (AI) are the key tools for the digital transformation in our daily life aspects [31, 32]. Machine learning and artificial intelligence (AI) can boost sustainable energy production. Sustainable energy is the core solution against climate change. A low carbon future is essential for humans and their living on Earth. Sustainable energy such as solar and wind energies are one of the key renewable and clean resources that can be applied to enhance several industrial activities. Moreover, wind and solar energy are a carbon-free technology that can help to encounter global climate change. Wind and solar power stations can collect big data dimensions for numerous metrological variables such as temperature, humidity, air pressure, wind direction, wind speed, and turbine power. Accordingly, the collected data have a large variety, huge volume, and high velocity. Therefore, these collected big data can be converted into useful information to manage the wind and solar stations effectively. Obtaining accurate wind speed and solar power predictions plays a decisive role in ensuring the power system's reliable operation is integrated with large-scale wind or solar power. Therefore, precise machine learning algorithms can be applied in the wind and solar energy applications as follows: wind and solar power prediction [33], online capacity factor estimation [34], wind turbines selection [35], and early faults identification and online power monitoring [16].

On the other hand, long-term policies against climate change aim to make our plant completely sustainable against CO_2 emissions. These policies are 100% renewable energy production [36], zero energy building [37], smart green cities [38], and electric vehicles [39]. However, these policies need more time and technology to be constructed in our natural world. Therefore, the future trend of both science and Industry are directed to maintain these policies.

5   Conclusion

The Coronavirus (or COVID-19) is a serious global issue. Currently, the global death toll has reached over +930000 deaths and will undoubtedly continue to rise. In contrast, an environmental problem like air pollution, which has been estimated to cause 4.5–7 million premature deaths every year, rarely makes headlines. Therefore, instead of adding to the barrage of coronavirus coverage, the objective is to use the global response to COVID-19 as a tool and simulation to understand the best approach to spurring immediate climate action. In April 2020, daily global CO_2 emissions were decreased by −17% compared with the mean 2019 levels due to COVID-19 confinement. This result reflects the impact of COVID-19 policies against climate change and CO_2 emissions.

A zero-carbon future is one of the key objectives of the United Nations' goals. The more effective path against climate change and CO_2 emissions is moving towards more energy efficiency and less carbon per economic activity unit. Moreover, artificial intelligence (AI) can boost the value of renewable energy and sustainable homes and city construction. Therefore, this study suggests Sustainable Climate Change Policies (SCCPs) that depend on the COVID-19 policies, sustainability policies, and artificial intelligence (AI). The guidelines have been divided into short-term policies and long-term policies to fight climate change. The response to the Coronavirus demonstrates that the planned economic slowdowns policies are necessary and effective to cut emissions drastically to avoid climate change and global warming.

References

1. Gleditsch, N.P.: This time is different! Or is it? NeoMalthusians and environmental optimists in the age of climate change. J. Peace Res. 0022343320969785 (2021)
2. Vardy, M., Oppenheimer, M., Dubash, N.K., O'Reilly, J., Jamieson, D.: The intergovernmental panel on climate change: challenges and opportunities. Annual Rev. Environ. Res. 17(42), 55–75 (2017)

3. Johnson, S.S., Constible, J., Knowlton, K., Gifford, B., Roberts, J.D., Ada, M.S., Jette, S.L.: Knowing well, being well: well-being born of understanding: Climate change & well-being: the role for health promotion professionals (2021)
4. Oldekop, J.A., Horner, R., Hulme, D., Adhikari, R., Agarwal, B., Alford, M., Bakewell, O., Banks, N., Barrientos, S., Bastia, T., Bebbington, A.J.: COVID-19 and the case for global development. World Develop. 1(134), (2020)
5. Lu, Y., Nakicenovic, N., Visbeck, M., Stevance, A.S.: Policy: five priorities for the UN sustainable development goals. Nature 520(7548), 432–433 (2015)
6. Sharma, V.K., Jinadatha, C., Lichtfouse, E.: Environmental chemistry is most relevant to study coronavirus pandemic (2020)
7. Norouzi, N., de Rubens, G.Z., Choubanpishehzafar, S., Enevoldsen, P.: When pandemics impact economies and climate change: Exploring the impacts of COVID-19 on oil and electricity demand in China. Energy Res. Social Sci. 1(68), (2020)
8. Rocklöv, J., Dubrow, R.: Climate change: an enduring challenge for vector-borne disease prevention and control. Nature Immunol. 21(5), 479–483 (2020)
9. Watson, M.F., Bacigalupe, G., Daneshpour, M., Han, W.J., Parra Cardona, R.: COVID-19 interconnectedness: health inequity, the climate crisis, and collective trauma. Family Process (2020). Accessed 26 June (2020)
10. Hammam, A.A., Elmousalami, H.H., Hassanien, A.E.: Stacking deep learning for early COVID-19 vision diagnosis. In: Big Data Analytics and Artificial Intelligence Against COVID-19: Innovation Vision and Approach (pp. 297–307). Springer, Cham (2020)
11. Kniffin, K.M., Narayanan, J., Anseel, F., Antonakis, J., Ashford, S.J., Bakker, A.B., Bamberger, P., Bapuji, H., Bhave, D.P., Choi, V.K., Creary, S.J.: COVID-19 and the workplace: implications, issues, and insights for future research and action (2020)
12. NASA GISS: Science Briefs: Greenhouse Gases: Refining the Role of Carbon Dioxide. http://www.giss.nasa.gov. Retrieved 26 April 2016
13. Kanniah, K.D., Zaman, N.A., Kaskaoutis, D.G., Latif, M.T.: COVID-19's impact on the atmospheric environment in the Southeast Asia region. Sci. Total Environ. 25, (2020)
14. Friedlingstein, P., Jones, M., O'sullivan, M., Andrew, R., Hauck, J., Peters, G., Peters, W., Pongratz, J., Sitch, S., Le Quéré, C., DBakker, O.: Global carbon budget 2019. Earth Syst. Sci. Data 11(4), 1783–838 (2019). Accessed 4 Dec 2019
15. Le Quéré, C., Jackson, R.B., Jones, M.W., Smith, A.J., Abernethy, S., Andrew, R.M., De-Gol, A.J., Willis, D.R., Shan, Y., Canadell, J.G., Friedlingstein, P.: Temporary reduction in daily global CO 2 emissions during the COVID-19 forced confinement. Nature Climate Change 19, 1–7 (2020)
16. Togelou, A., Sideratos, G., Hatziargyriou, N.D.: Wind power forecasting in the absence of historical data. IEEE Trans. Sustain. Energy 3(3), 416–421 (2012)
17. Elmousalami, H.H., Hassanien, A.E.: Day level forecasting for Coronavirus Disease (COVID-19) spread: analysis, modeling and recommendations. http://arxiv.org/abs/2003.07778 (2020). Accessed 15 March 2020
18. Janssens-Maenhout, G., Crippa, M., Guizzardi, D., Muntean, M., Schaaf, E., Dentener, F., Bergamaschi, P., Pagliari, V., Olivier, J.G., Peters, J.A., Van Aardenne, J.A.: EDGAR v4. 3.2 Global Atlas of the three major greenhouse gas emissions for the period 1970–2012. Earth Syst. Sci. Data 11(3), 959–1002 (2019)
19. Liu, Z., Deng, Z., Ciais, P., Lei, R., Davis, S.J., Feng, S., Zheng, B., Cui, D., Dou, X., He, P., Zhu, B.: COVID-19 causes record decline in global CO_2 emissions (2020)
20. Xiao, K.: Saving lives versus saving livelihoods: can big data technology solve the pandemic dilemma? Available at SSRN 3583919 (2020). Accessed 15 April 2020
21. Diffenbaugh, N.S., Field, C.B., Appel, E.A., Azevedo, I.L., Baldocchi, D.D., Burke, M., Burney, J.A., Ciais, P., Davis, S.J., Fiore, A.M., Fletcher, S.M.: The COVID-19 lockdowns: a window into the Earth System. Nature Rev. Earth Environ. 29, 1–2 (2020)
22. Rolnick, D., Donti, P.L., Kaack, L.H., Kochanski, K., Lacoste, A., Sankaran, K., Luccioni, A.: Tackling climate change with machine learning (2019). https://arxiv.org/pdf/1906.05433

23. Whyte, C.: Green new deal proposal includes free higher education and fair pay. Environment **12** (2019)
24. Djalante, R.: Key assessments from the IPCC special report on global warming of 1.5 °C and the implications for the Sendai framework for disaster risk reduction. Progr. Disaster Sci. **1**(100001) (2019). Accessed 1st May 2020
25. Ramachandra, T.V.: Renewable energy transition: perspective and challenges. Energy India. 175–183 (2020)
26. IPCC. Global warming of 1.5_C. An IPCC special report on the impacts of global warming of 1.5_C above pre-industrial levels and related global greenhouse gas emission pathways, in the context of strengthening the global response to the threat of climate change, sustainable development, and efforts to eradicate poverty In: Masson-Delmotte, V., Zhai, P., P¨ortner, H.O., Roberts, D., Skea, J., Shukla, P.R., Pirani, A., Chen, Y., Connors, S., Gomis, M., Lonnoy, E., Matthews, J.B.R., Moufouma-Okia, W., P´ean, C., Pidcock, R., Reay, N., Tignor, M., Waterfield, T., Zhou, X., (eds.) (2018)
27. Hamdy, M., Hasan, A., Siren, K.: A multi-stage optimization method for cost-optimal and nearly-zero-energy building solutions in line with the EPBD-recast 2010. Energy Build. **56**, 189–203 (2013)
28. Liu, Z., Liu, Y., He, B.J., Xu, W., Jin, G., Zhang, X.: Application and suitability analysis of the key technologies in nearly zero energy buildings in China. Renew. Sustain. Energy Rev. **101**, 329–345 (2019)
29. Boermans, T., Grözinger, J.: Economic effects of investing in EE in buildings–the BEAM2 Model. In Background paper for EC Workshop on Cohesion policy (2011)
30. Wesselink, B., Deng, Y.: Sectoral emission reduction potentials and economic costs for climate change (SERPEC-CC), summary report (2009)
31. Elmousalami, H.H. Comparison of artificial intelligence techniques for project conceptual cost prediction: a case study and comparative analysis. IEEE Trans. Eng. Manag. (2020a)
32. Elmousalami, H.H.: Artificial intelligence and parametric construction cost estimate modeling: state-of-the-art review. J. Constr. Eng. Manag. **146**(1), 03119008 (2020b)
33. Di Piazza, A., Di Piazza, M.C., La Tona, G., Luna, M.: An artificial neural network-based forecasting model of energy-related time series for electrical grid management. Math. Comput. Simulat. (2020). Accessed 16 May 2020
34. An, X., Jiang, D., Zhao, M., Liu, C.: Short-term prediction of wind power using EMD and chaotic theory. Commun. Nonlinear Sci. Numer. Simulat. **17**, 1036–1042 (2012)
35. Sideratos, G., Hatziargyriou, N.D.: Probabilistic wind power forecasting using radial basis function neural networks. IEEE Trans. Power Syst. 1–9 (2012)
36. Lund, H., Mathiesen, B.V.: Energy system analysis of 100% renewable energy systems—The case of Denmark in years 2030 and 2050. Energy **34**(5), 524–531 (2009)
37. Marszal, A.J., Heiselberg, P., Bourrelle, J.S., Musall, E., Voss, K., Sartori, I., Napolitano, A.: Zero energy building–a review of definitions and calculation methodologies. Energy Buildings **43**(4), 971–979 (2011)
38. Antrobus, D.: Smart green cities: from modernization to resilience? Urban Res. Pract. **4**(2), 207–214 (2011)
39. Chen, Z., He, F., Yin, Y.: Optimal deployment of charging lanes for electric vehicles in transportation networks. Trans. Res. Part B Methodol. **1**(91), 344–365 (2016)
40. Metcalf, G.E., Weisbach, D.: The design of a carbon tax. Harv. Envtl. L. Rev. **33**, 499 (2009)

The Correlation Between Human Lifestyle, Weather, Smart Technologies and Covid-19 Pandemic

Yasmine S. Moemen, Ibrahim El-Tantawy El-Sayed, Ashraf Darwish, and Aboul Ella Hassanien

Abstract The SARS-CoV-2 epidemic has caused numerous deaths and injuries globally, striking all communities. Till this moment, a health hazard will be continued for years. The outbreak is a chance for the polity to reevaluate the use of housing, transportation, social housing in helping individuals, especially susceptible people, and extra green public spaces to support a healthier atmosphere that will define how differences outline the exposure, susceptibility; besides, the risk and consequence of infectious diseases. Increasing community and environmental vulnerabilities will be accompanied by COVID-19 that drastically increased the infection risk and death. During COVID-19, emerging technology such as drones and smart transportation can play a crucial role in fighting this virus. Besides, there is an impact of climate change on the spreading of COVID-19.

1 Introduction

The Severe Acute Respiratory Syndrome Coronavirus 2 (SARS-CoV-2) or Coronavirus Disease 2019 (COVID-19) is responsible for about 44,710,816 infections and 1,178,021 deaths worldwide; while 1,759,447 infections and 42,201

Y. S. Moemen (✉)
Clinical Pathology Department, National Liver Institute, Menoufia University, Menoufia, Egypt
e-mail: yasmine_moemen@liver.menofia.edu.eg
URL: http://www.egyptscience.net

I. E.-T. El-Sayed
Chemistry Department, Faculty of Science, Menoufia University, Menoufia, Egypt

A. Darwish
Faculty of Science, Helwan University, Helwan, Egypt

A. E. Hassanien
Faculty of Computer and Artificial Intelligence, Cairo University, Cairo, Egypt

Y. S. Moemen · A. Darwish · A. E. Hassanien
Scientific Research Group in Egypt, Cairo, Egypt

© The Author(s), under exclusive license to Springer Nature Switzerland AG 2021
A. E. Hassanien et al. (eds.), *The Global Environmental Effects During and Beyond COVID-19*, Studies in Systems, Decision and Control 369,
https://doi.org/10.1007/978-3-030-72933-2_9

137

deaths in Africa; besides, 107,030 thousand infections and more than 1,451 of deaths in Egypt [1]. It is a viral droplets come from several viruses, influenza and few coronaviruses usually start in cold weather and end in warm weather. Although COVID-19 still exists because people regularly contact inside places with insufficient air, which will increase the risk of infection, as told by Mauricio Santillana, a mathematician at Harvard Medical School in Boston, Massachusetts, who predicted the current pandamic [2].

Most individuals infected with SARS-CoV-2 symptoms will have mild to moderate respiratory disease that might be recovered without physician consultation. Individuals with chronic diseases like cardiovascular, diabetes, chronic respiratory disease and cancer might develop acute respiratory symptoms. The prevention of COVID-19 transmission should include keeping social distance among people, washing hands or using alcohol and not touching the face. It transmits through droplets of saliva from the person when an infected individual coughs or sneezes [1]. At this time, there is no available vaccine or therapy for COVID-19. However, many current clinical trials are evaluating potential drugs.

The understanding of human demography, besides the dynamic factors of weather [3], vulnerable reduction and interference; where, will define the epidemic spread; so more active responses will be prepared. There is much human demography in the United States (US), residents' mass, environmental influences, much available data and a respect for the social distance interferences, once interventions are described [3]. A spotlight on diverse urban areas advises that some places benefited from the control of a nearly immediate worldwide closure and the local weather environments in mid-March, while others suffer significant epidemics. Predictions for decreasing the spread of current pandemic disease can be developed by the first-guidelines type of the infection spread, defining proper social distancing and regional weather requirements. A limited area, as a specific portion of people, came sick and exhibited signs where a pandemic has partly defined by excluding vulnerable people, herd immunity. At the same time, most regions' pain of COVID-19 spread [3].

2   Methods of COVID-19 Confining

The plague is a chance for governments to reconsider dense population, transportation and social housing to serve people, especially the socially exposed. Besides, increasing green community spaces in urban and cities towards a healthier environment that will demonstrate how differences outline the exposure, susceptibility, ultimately the risk and outcome of infectious diseases. Increasing community and environmental vulnerabilities will be accompanied by COVID-19 that drastically increased the infection risk and death [4].

2.1 Population Density

Regardless of thoughts that metropolitan mass exaggerates outbreaks like COVID-19, this is due to home congestion and unsafe lodging conditions besides the socio-spatial imbalances. Also, low minor populations are frequently at risk of more surrounding environment hazards, fewer environmental facilities, health decreases, and less life expectancy than white people. Simultaneously, rich people stay in wonderful houses with many living spaces, which will decrease the danger of virus spread [4].

In 2020 spring, the United Kingdom, the five highest density populations, showed 70% more infected coronavirus people than the five least density populations. To stop the pandemics, cities need reasonable, acceptable, safe and available housing. Governments should provide a moratorium, loans, and exclusions for vulnerable people [4].

2.2 Transportation and Smart Transportation During COVID-19

Public transport systems are broadly respected as transmission hotspots. It is cheap, crowded and may take a long time so, many proficient specialists can work remotely to maintain a social distance from journeying on such organizations and the richest who likely turn to private modes of transport. So, poor laborers cannot find any alternative system for public transportation which will put them at high risk of COVID-19 [4].

The current epidemic is a worldwide challenge due to its high-level transmission and our hypermobile society. Besides, there is no definite health defense like vaccines or treatment that can contain the pandemic globally. The transport division, whether in rail, water and air transport have all been influenced. Traveler and shipping transport has also been jammed harshly due to the dense resources and requirement developments. The decrease in transport need, therefore, translates into a decline in energy and fuel request. During COVID-19, the overall electricity demand amidst pandemic conditions has been declined. An intelligent town represents efficient mobility; more technology use in planning and management that helps citizens screen the town quickly, timely and comfortably. The incorporation of knowledge and ICT is crucial to any city's wealth and the enhancement to mobility division. It refers to the increase modern technologies use in transport division for new towns. Novel technologies are motive for quick development, improve several facilities and leading to novel town competitive globally. Additionally, novel technologies like developed resources, artificial intelligence, cyber resilience, space systems, remotely piloted techniques and independent technologies. Mobility and transportation are one of the

main features of a city. The fact that all services are interconnected and heavily populated inspired cities to develop various transportation solutions. It involves useful
urban development and easy transportation solutions in the environment [5, 6].

2.3 Community Places

Current epidemic presents towns' chance for using open spaces like sidewalks or cars
broader, bike paths and free streets. However, the car makers can force novel political
measures versus community. Besides, community leaders are mindful of COVID-
19, which can put 1.1 m vehicle fabricating employment at risk within the European Union alone. Cities got to move quickly on the off chance to reconfigure
roads as community spaces and face the car makers' exploit.

Currently, cleaning lanes is an urgent need as contamina-
tion causes chronic diseases like heart and lung damage, which will worsen COVID-
19 patients, regain pedestrian rights and help secure post-COVID towns in
case of disease and crashes. Moving to healthy towns is a genuine attempt to
form towns fresher and impartially green. In Valencia, Spain, and Nantes,
France, decentralized systems of little green spaces are giving inhabitants with simple nature for all inhabitants without compromising get to bigger parks.
Many cities should consider expanded utilize of empty spaces such as flat rooftops
that can be changed over into community gardens and give more green space.

2.4 Managing Priorities

Three major urban areas form the structure that urban variations can decrease
spreading discriminations. Social discrimination periods have caused low income
besides minor societies to become more health risk and financial weakness. There
is an additional load of the COVID-19 disease and its financial penalties. There is
a need to change such three areas that are more significant in the global South; the
community justice values are essential, although feedback must be deep-rooted in
government law and priorities. To avoid expanding pandemics, societies, cities and
their supporting inadequate financial constructions must be changed to justice as the
current organization should be fixed, supported and upgraded to serve all people,
specifically the socially susceptible to Covid-19.

3   Results and Discussion

3.1   Population Density

The main message evolved from the worldwide evaluation is that more healthy environments can prohibit early deaths and ailments. Investigating the up-to-date information on the environment disease links the shocking effect of environmental threats on worldwide health and is supported by professional judgment; where, description demonstrates further 100 diseases and injuries. In 2016, the World Health Organization (WHO) reported that 24% of worldwide mortality and 28% of these cases in children under five are due to environmental alternation issues. 68% of mortalities and 51% of disease burden can be assessed with rational danger evaluation approaches. Extra epidemiologic evaluations and skillful judgment finished the calculations of other environmental exposures. Ischemic heart disease, chronic respiratory diseases, cancers and accidental injuries head happen to Individuals in low- and middle-income nations tolerate the extreme disease load as listed in supplementary Tables in 2016 and updated to the year 2019 [7] and as shown in Figs. 1, 2, 3.

These calculations enhance motivation to manage worldwide efforts for inspiring healthy environments. This study will notify about the Sustainable Development Goals' transformational spirit agreed by Heads of State in September 2015. The consequences of such a study highlight the demand for robust department act to generate healthier environments that will donate to the long-lasting development of millions of lives worldwide.

Population in High-Income Countries (HIC), Population in Low and Middle-Income Countries (LMIC) were shown in Fig. 1. At the same time, Deaths and disability-Adjusted Life Years (DALYs), Total environment and burden attributable to environment displayed in Fig. 2, and non-communicable diseases, region were collected and updated by WHO for HIC countries status, 2019, demonstrated in Fig. 3.

4   Technologies to Improve Quality of Life During the Pandemic

4.1   Drones in the Era of COVID-19

This system is used to monitor and control COVID-19 activities remotely. Here, a set of smart devices is interconnected for sharing the necessary information between various parties. After that, it is used for various other purposes, such as making preparations for purchasing medical and other types of equipment and medicines. COVID-19 pandemics and hotspots can be monitored remotely using this system. Further, statistics collected from drones can be shared in the public domain for

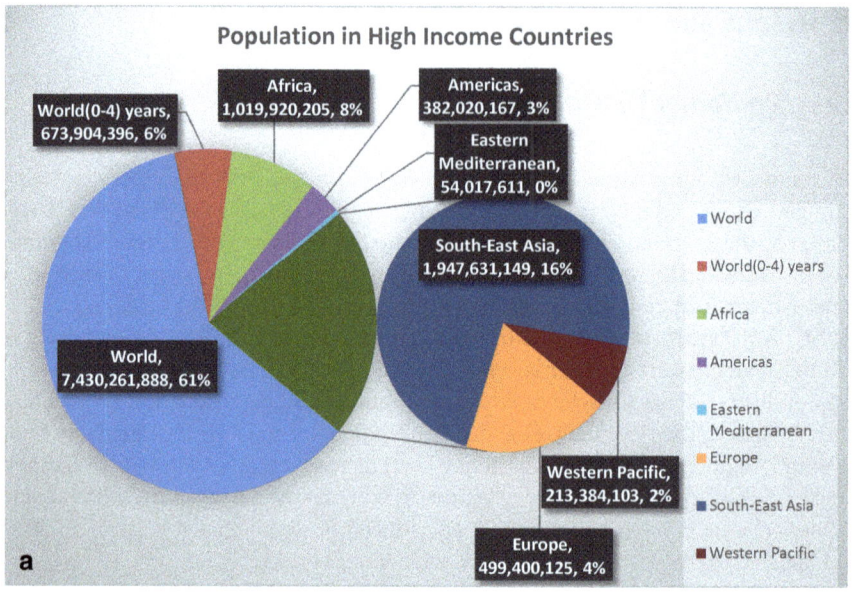

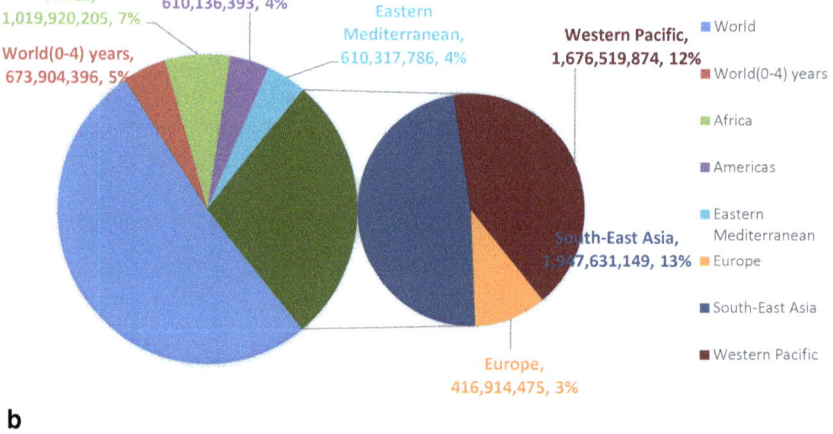

Fig. 1 **a** Population in low-income countries. **b** Population in the middle- and high-income countries

creating awareness and protection. This system helps the government and non-profit Non-Governmental Organizations (NGOs) to extend their help for patients living in COVID-19 affected areas and hotspots [8].

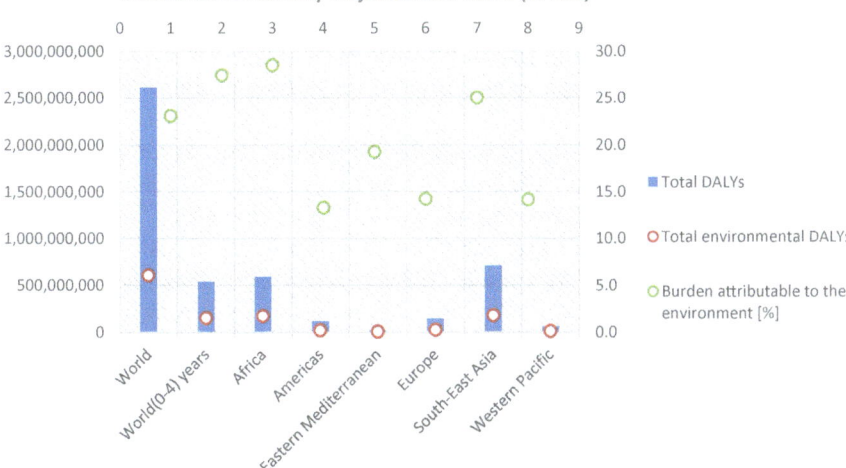

Fig. 2 Deaths and disability-adjusted Life Years (DALYs), total environment and burden attributable to the environment

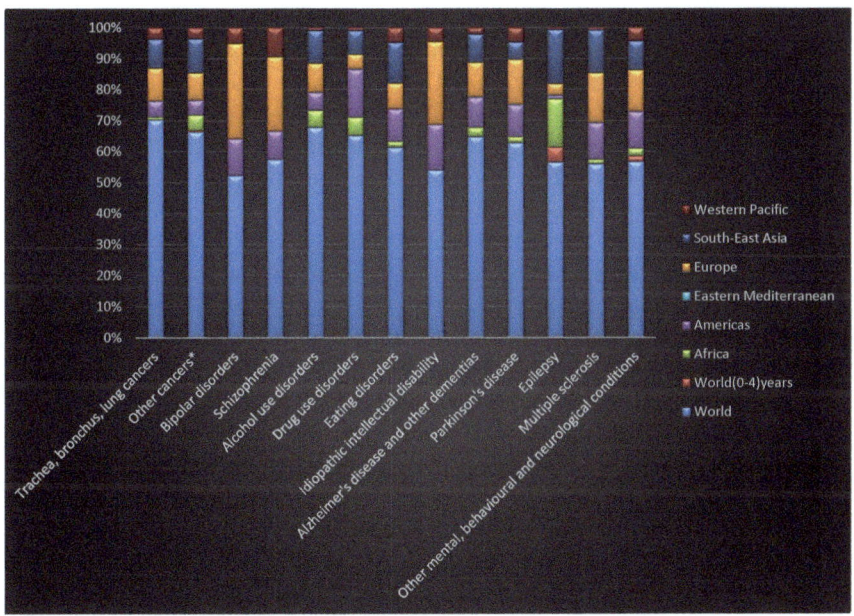

Fig. 3 Non-communicable diseases, region have been collected and updated by WHO for HIC countries status, 2019

4.1.1 Drone-Network and Data Systems

1. Drone Monitoring Centre: This sub-system monitors drone activities. The drone device performance, movements, services and operations are observed in this sub-system.
2. Drone Control entre: In this sub-system, controlling the party instructing the monitoring party to move the drone in the area of COVID-19 patients and start the necessary operations.
3. Data Collection Centre: This sub-system focused on the types of data to be collected from COVID-19 affected areas and prepared a well-defined template. This template and the data are forwarded for necessary processing and decision-making.

4.1.2 Smart Healthcare System

1. **Patient Thermal Image Record**: This is the essential record for medico and non-medico persons associated with COVID-19 operations. The thermal images reflect the temperature or density-based movements that need sanitization regularly.
2. **Patient Health Record**: The patient's health record is essential to consider when a person is affected by COVID-19 or similar viruses. Thus, medication needs patient history that is possible through maintain historical records. This is made feasible through the patient health record sub-system.
3. **Patient Medical Record**: In this sub-system, the patient's medical records are maintained. The medical records are different from health records in terms of specialized data. In medical records, disease-specific records are maintained. Whereas, health record means the general medical practices and routine followed to maintain the health.
4. **Patient Personnel Record**: Like health and medical records, patient's personnel records are essential as well. These records are necessary to have personal identification and authentic payment sources for treatment.
5. **Patient COVID-19 Record**: This record is recently found to be necessary for finding the chain of airborne diseases like COVID-19. Individual patient's symptoms, treatments and cures are considered as single-clinical trials. Thus, it helps identify the right treatment when data is significant, and it is properly analyzed.
6. **Isolation and Quarantine System**: This system takes care of activities required when patients are affected or found to be COVID-19 positive case. Thus, processes and care are defined, monitored, and practiced in this sub-system.

4.2 Climate Effect During COVID-19

The seasonal outcome is the main reason for the enlarged spread of COVID-19. Many individuals are vulnerable to the risk of infection, says Rachel Baker, an epidemiologist at Princeton University in New Jersey [2]. So people in warm weather should be worried.

The main reason that affects the pandemic proportions can be managed procedures like social distancing and using a mask. Periodic movements in infection by the virus are driven by many parameters, like actions of people and the attributes of the virus, which can't live in warm and moist environments [9].

Research laboratory tests expose that SARS-CoV-2 preferer cold, dry conditions and avoiding sunlight. As non-natural ultraviolet radiation could kill SARS-CoV-2 molecules in places [10] and on-air [11]; particularly, in temperatures near 40 °C. Virus infectious also damaged quicker on warm surfaces and wet conditions [12]. "In wintertime, individuals preferred to warm their houses near 20 °C, dried air", says Dylan Morris, a mathematical biologist at Princeton. "Indoor conditions in the winter are pretty favorable to viral stability" [2]. To estimate the risk of SARS-CoV-2 infections, a classical academics study reported that the virus spreads in a certain place, several times a year and over different years. They also studied the climate effect on SARS-CoV-2 spread by observing infection rates in several areas globally [2].

A research study [13] observed the infection's development in the early epidemic months, before worldwide imposing infection control procedures. It concluded that diseases spread in places with low UV light; so, infected people would decrease in warm weather and increase in cold weather. In the cold, "the risk goes up, but you can still dramatically reduce your risk by good personal behavior," says Cory Merow, an ecologist at the University of Connecticut in Storrs and a co-author of the study. "The weather is a small drop in the pan" [2].

But Francois Cohen, an environmental economist at the University of Barcelona in Spain, says, "COVID-19 testing was restricted early in the plague and continued to be variable, so it is intolerable to control the effect of climate on the spread of the virus so far".

Baker has tried to take part in the effect of climate on the seasonal case pattern during the pandemic, using information about another coronavirus's sensitivity to moisture [2]. She demonstrated [14] the increase and decrease infection risks over several years for New York City with and without a weather effect and with diverse levels of procedures control. They found that a minor climate effect can result in significant outbreaks when the seasons vary if control procedures are only just handled to confine the virus. "That could be a location where climate might nudge you over," Baker says. The team posted its results on the preprint server medRxiv on 10 September; the authors suggest that severe control procedures might be wanted through winter to decrease plague hazards.

Display the relation between the COVID-19 pandemic, climate change, the range of infections spread and warmer weather effects on the spread of coronavirus. The relation between climate changes and the transmission of coronavirus. The impact of early warning systems and the impact of COVID-19 in the agricultural sector. Food demand impacts are introduced, agriculture research and COVID-19 pandemic impacts. Finally, the efforts to meet the food demands [15].

Climate change affects not only food production, but it affects all the four food security dimensions as follows:

1. Availability: loss in food production and indirect environmental feedback.
2. Access: infrastructure damage, losses of asset, income and employment opportunities.
3. Stability: increased livelihood risks, raising food prices, dependency on food imports, and food aid.
4. Utilization: human health risks and nutrition.

4.3   Conclusion and Future Trends

If SARS-CoV-2 endure cold environments, so it will hard to separate participation from the consequence of people's actions. Flu has been existing for decades and, in certain ways to has peaks of flu in the winter, which is not fully understood. Even if researchers had more accurate data for SARS-CoV-2, they would see only minor or negligible seasonal effects so early in the pandemic, when much of the population is still susceptible. Over time, however, climate effects could play a more critical part in driving infection trends as more people build up immunity to the virus. It could take up to five years through natural infection or less if people are vaccinated. The seasonal pattern will depend on several issues that will not be fully understood, that includes how long immunity herd, how long healing takes time and how likely it is that people can be reinfected.

References

1. World Health Organization.: WHO Coronavirus Disease (COVID-19). https://www.worldomet ers.info/coronavirus/#countries (2020). Accessed 22 Oct 2020
2. Mallapaty, S.: Why COVID outbreaks look set to worsen this winter. https://www.nature.com/articles/d41586-020-02972-4 (2020). Accessed 23 Oct 2020
3. Araujo, M.B., Naimi, B.: Spread of SARS-CoV-2 Coronavirus likely to be constrained by climate. medRxiv (2020)
4. Anguelovski, I.: COVID-19 highlights three pathways to achieve urban health and environmental justice. https://www.iied.org/covid-19-highlights-three-pathways-achieve-urban-hea lth-environmental-justice (2020)
5. Abu-Rayash, A., Dincer, I.: Analysis of mobility trends during the COVID-19 coronavirus pandemic: exploring the impacts on global aviation and travel in selected cities. Energy Res. Soc. Sci. **68**, (2020)
6. Sung, J., Monschauer, Y.: Changes in transport behaviour during the Covid-19 crisis. What can we learn from the lessons of the past?. https://www.iea.org/articles/changes-in-transport-beh aviour-during-the-covid-19-crisis (2020). Accessed 10 Dec 2020

7. Updated 2016 data tables for 'Preventing disease through healthy environments'.: Geneva: World Health Organization, 2019 (WHO reference number). Licence: CC BY-NC-SA 3.0 IGO. https://creativecommons.org/licenses/by-nc-sa/3.0/igo (2019)
8. Kumar, A., Sharma, K., Singh, H., Naugriya, S.G., Gill, S.S., Buyya, R.: A drone-based networked system and methods for combating coronavirus disease (COVID-19) pandemic. Futur. Gener. Comput. Syst. **115**, 1–19 (2020)
9. Afshordi, N., Holder, B., Bahrami, M., Lichtblau, D.: Diverse local epidemics reveal the distinct effects of population density, demographics, climate, depletion of susceptibles, and intervention in the first wave of COVID-19 in the United States. http://arxiv.org/abs/2007.00159 (2020)
10. Ratnesar-Shumate, S., et al.: Simulated sunlight rapidly inactivates SARS-CoV-2 on surfaces. J. Infect. Dis. (2020)
11. Dabisch, P., et al.: The Influence of Temperature, Humidity, and Simulated Sunlight on the Infectivity of SARS-CoV-2 in Aerosols. Aerosol Sci. Technol. 1–15 (2020)
12. Riddell, S., Goldie, S., Hill, A., Eagles, D., Drew, T.W.: The effect of temperature on persistence of SARS-CoV-2 on common surfaces. Virol. J. **17**(1), 1–7 (2020)
13. Merow, C., Urban, M.C.: Seasonality and uncertainty in global COVID-19 growth rates. Proc. Natl. Acad. Sci. (2020)
14. Baker, R.E., Yang, W., Vecchi, G.A., Metcalf, C.J.E., Grenfell, B.T.: Assessing the influence of climate on future wintertime SARS-CoV-2 outbreaks. medRxiv (2020)
15. Manzanedo, R.D., Manning, P.: COVID-19: Lessons for the climate change emergency. Sci. Total Environ. **742**, (2020). https://doi.org/10.1016/j.scitotenv.2020.140563

Exploring the Impacts of COVID-19 on Oil and Electricity Industry

Atrab A. Abd El-Aziz, Nour Eldeen M. Khalifa, and Aboul Ella Hassanien

Abstract The COVID-19 Pandemic has dramatically influenced the global market of energy production and consumption. This influence could be noticed obviously by the latest drop in crude oil prices. Furthermore, coronavirus has affected the supply chains and delayed the development of sustainable energy worldwide. Due to its relevance, academics have begun to study the associations regarding this crisis. The COVID-19 Pandemic offers a new chance to investigate the impacts of prolonged landscape-scale confusion on sustainability change paths in real-time. How the global renewable energy flexibility sector will respond to the COVID-19 Pandemic is a critical question. This crisis could inspire governance structures to plan adequately for other varieties of crises in the future. These improvements can drive research by spouting the generation of new disciplines stemming from the COVID-19 outbreak to expedite sustainability transitions and improve the recognition of governance's main role in changes. Smart policies could transform COVID-19 threats into great opportunities for the world's sustainable energy scenario towards green energy generation and use in the coming years. In this paper, the impacts of COVID-19 in terms of the energy sector, especially the electricity and oil sectors, will be explained. The major objective of this research is to shed light on future research on renewable energy.

Keywords Energy · COVID-19 · Green energy · Smart policies · Oil · Electricity

A. A. Abd El-Aziz
Faculty of Computers and Information, KafrelSheikh University, Kafr El Shaikh, Egypt
e-mail: Atrab_Ahmed@fci.kfs.edu.eg

N. E. M. Khalifa (✉) · A. E. Hassanien (✉)
Faculty of Computers and Artificial Intelligence, Cairo University, Cairo, Egypt
e-mail: aboitcairo@cu.edu.eg
URL: http://www.egyptscience.net

A. A. Abd El-Aziz · A. E. Hassanien
Scientific Research Group in Egypt (SRGE), Cairo, Egypt

1   Introduction

The World Health Organization has announced COVID- 19 to be a global pandemic in December 2019 [1]. The outbreak of COVID-19 has direct and indirect effects, The COVID-19 epidemic has direct and indirect consequences, which are likely to be even more frequent over time. Economic growth in various countries is slowing as principal financial and manufacturing markets are experiencing a substantial decline, foreign supply chains are breaking, borders are closing, and tourism is slowing down. The COVID-19 outbreak provides a rare opportunity to examine the impact of a sustained landscape-scale disturbance on real-time trajectories of biodiversity change [2].

Energy can satisfy basic human needs. Humans do not require energy, but the use of the services it offers, such as lighting. Exposure to these energy resources is generally provided as a consumer product, ensuring that people can use electricity as much as they can afford it. The principle of energy sovereignty encompasses self-determination rights, which individuals and societies possess legal and natural rights. This helps on the decisions-making-process about the sources, sizes, modes of ownership, and the access which constitute the energy systems employed for the provision of energy services [3]. The COVID-19 Pandemic hit renewable energy production facilities, businesses, and supply chains and reduced the transition to renewable energy sources. Even entrenched renewable energy policies are in doubt, especially those that weigh heavily in industries severely affected by the crisis. Many countries' budgets will indeed be adjusted, and extra renewable energy programs will surely be delayed. Construction programs for companies installing equipment for renewable energy technologies will be delayed amid austerity measures. The global renewable energy outline that flourished in modern decades, and owned rapid increase, faced a severe challenge due to the coronavirus.COVID-19 has declined major oil producers [4] and has lowered natural gas prices to two dollars (Fig. 1).

As a consequence, gas will soon be the residual coal power plant in Europe [5]. The reduction in fossil fuel prices is particularly problematic in developing countries where low-cost electricity supplies seem necessary because of their vulnerable economic status at COVID-19. Its high energy cost sensitivity will push policymakers to embrace cheaper traditional sources of energy rather than renewable energy, which would be bad for global climate policy.

Banks that manage very low-interest rates will escape this uncomfortable situation in the face of an economic crisis that threatens the introduction of high-cost renewable energy programs that will restrict the energy market from shifting towards energy production from fossil fuels. The COVID-19 Pandemic is recognized as a health care emergency and a global economic and financial emergency. Energy plays an essential role in the formation of the countries' economies [6]. COVID-19 has revealed the weakness of the continent in the energy sector. It has also revealed the strength and resistance of companies and infrastructure throughout the world. Also, it is a tragedy once the Pandemic is rapidly spreading. Many portions are responsible for energy consumption in buildings for the domestic sector [7].

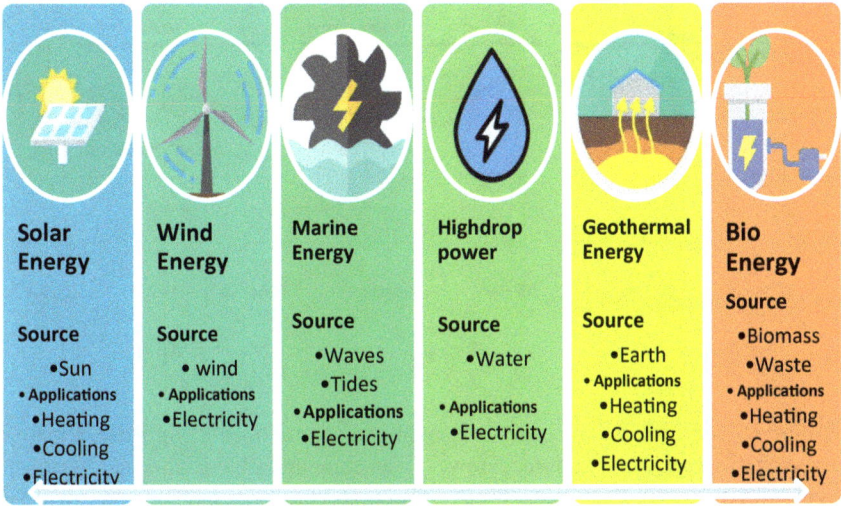

Fig. 1 Sources of renewable energy, applications and technologies. *Source* ECA

This paper reviews the impact of COVID-19 on energy sector dynamics. The Pandemic has unforeseen implications that lead to some of these and other problems while providing opportunities to promote access to energy through renewable energy technologies. Accordingly, this document analyzes these main challenges and proposes policy directions to overcome these challenges by taking advantage of the available opportunities.

2 Electricity Consumption Shift During Coronavirus

In March 2020, many European countries implemented quarantine like measures led by Italy. The European governments' procedures to fight the COVID-19 crisis have altered people's activities and habits at the state level. These behavior changes are expressed in electrical systems, especially in terms of changes in electricity use profiles.

Europe's electricity generation began to decline rapidly, particularly in the most affected countries by COVID-19. In early 2020, electricity production in Europe followed a trend similar to that seen in 2016-2019. However, a sharp decline is reported as the number of confirmed COVID-19 cases increased and countries took measures similar to those observed. While year-on-year changes in power generation and use may be due to seasonal conditions changes, the exceptional changes observed can be primarily attributed to restrictions imposed to contain the spread of COVID-19 [8].

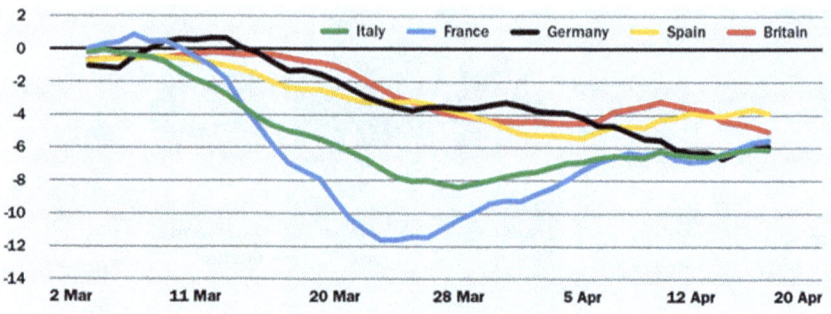

Fig. 2 Electricity demand reductions after lockdown moves, from March to April

There was an increase in domestic demand during the lockdown because people were spending extra time at home. However, as companies have reduced their operations, the fall in industrial and commercial demand has been much greater than the increase in domestic demand. In June, electricity demand in most nations except India, where the recovery is more pronounced, corrected for the weather, remained 10% below what it was before the first lockdown measures. However, with the containment measures, the electricity demand dropped rapidly, as shown in Fig. 2. This has gradually increased as the steps have grown weaker and weaker, still 10% below the EU countries' pre-lock rate [9].

3   OIL Demand During COVID-19 Pandemic

The impact of COVID-19 has been severe in oil-dependent countries. The global fall in oil prices, together with the reduction in demand for petroleum products in the international market, has led to a significant shortage of oil revenues for oil-dependent countries. This led to an increase in the deficit in the current account and exacerbated the balance of many oil-dependent countries' payments situation. Most of the oil demand comes from the transportation sector, with about 50% after industry consumption. Oil prices fell as a result of Russia's oil price war with Saudi Arabia in early 2020. Also, the coronavirus pandemic made the situation worse by reducing oil demand. This resulted in a drop in demand for coal, aviation fuel, and extra energy products, which resulted in a subsequent reduction in oil prices because of low demand. The Pandemic's effect on the oil markets was more severe as it impeded people and products' movement. This culminated in a dramatic decrease in demand for transportation fuels. Another reason for the drop in oil price occurred when Saudi Arabia supplied the world with surplus oil and saturated the market with too much crude.

This led to crushing demand during the COVID-19 Pandemic and ultimately contribute to a collapse in oil prices [10].

On April 20, 2020, WTI (West Texas Intermediate) prices collapsed for many reasons, as shown in Fig. 3. The effect of ongoing, on-demand development of COVID-19 in the battle against storage limitations is one of them. But the key explanation was that the contracts for the WTI oil outlook were completed on April 21 and the US Petroleum Fund owned 25% of the WTI futures for May 2020.

At the end of May, they had to receive a large quantity of oil, knowing the situation of reduced storage, or sell it instantly at the price they could obtain for it. Many countries were severely affected during the coronavirus pandemic by consuming basic products, such as the USA and Canada.

A. United States of America (USA)

In the United States of America, shale oil production has exploded since 2014. As of early 2020, this represents more than a third of onshore crude oil production in 48 states. As a result, the United States of America became the world's largest crude oil producer, and its crude imports decreased by 15% between 2013 and 2019. Between November 2019 and April 2020, over 300 drilling rigs were closed in the time of a pandemic outbreak. US gas and oil companies have reduced their capital expenditures by nearly $ 100 billion in the last months. This led the United States of America to import heavy crude oils to meet domestic demand and American refineries' configurations. This will help some heavy stock producers. However, the main concern will remain the economic viability of producing shale oil [11].

B. Canada

Canada has several relatively light crude oil deposits, mostly concentrated in Alberta's tar sands and API gravity deposits ranging from 19 to 22. Despite the Pandemic in demand and falling prices to the point that it is cheaper than the cost

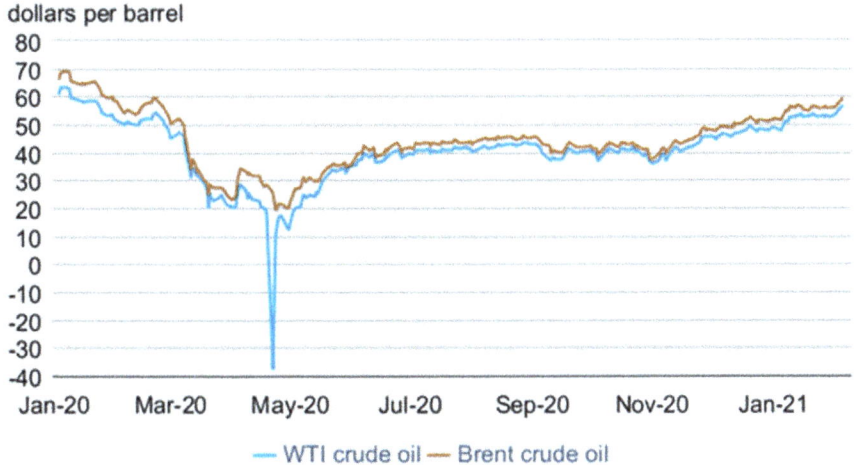

Fig. 3 Weekly WTI crude oil prices from December, 2019 to September, 2020. *Source* CME

of exporting it to the already poor American economy, at just $5 a barrel. Although closing oil sands is difficult and expensive, there have been reports of closures of up to 25% [12]. By the end of March 2020,Canada's economy was already on the verge of collapse. The heart of Canada's oil industry is in Alberta and Saskatchewan, where up to 200,000 job declines are expected. [13].

4 Impacts of COVID-19 on the Energy Sector

The energy sector is segmented into two parts, the traditional energy sector and the new energy sector [14].

Traditional energy: refers to using and developing various natural energy resources and then transforming them into secondary energy. Production includes energy, mining, power industry chains, natural gas, oil, and coal smelting.

New energy: involves the utilization and development of some modern sources. It is essentially classified into two classes: (a) the new energy such as wind energy, water energy, nuclear energy, solar energy, and biomass energy. (b) The new driving force formed by technological innovations. There are significant differences in manufacturing processes and sources of raw materials between them, and substantial investments in fixed assets recognize them and therefore have high fixed costs.

Due to the COVID-19 outbreak, the government imposed strict control on production and closed factories even in high-risk areas, resulting in lower corporate income. Below the double pressure to reduce revenue and increase costs, its uncertainty is greatly improved. Therefore, the uncertainty caused by COVID-19 makes investors, especially creditors and banks, more risk-averse. Listed companies will face greater financial constraints, which will lead to lower operating cash flows and lower performance. Figure 3 shows the various aspects that are affected by COVID-19 (Fig. 4).

4.1 *Energy Consumption*

In Pandemic, a study conducted in New York City revealed that total energy consumption in the commercial and industrial sector has decreased by about 7%. On the other hand, consumption grew in March to almost 23% and to 10% in April 2020 for domestic households [15]. The curfew at the time of the Pandemic has led to an increase in electricity bills. This increase with a loss of income resulted in a huge financial and economic burden on households. Government suspension of utility bills in times of Pandemic mitigates the financial responsibility of households. Therefore, current governments have a crucial need for a policy plan to address the COVID-19 pandemic situation economically, as shown in Fig. 5.

For certain parts of the world, the short-term period is known as the closure era, has been ended. The medium-term can be described as the time of social distancing

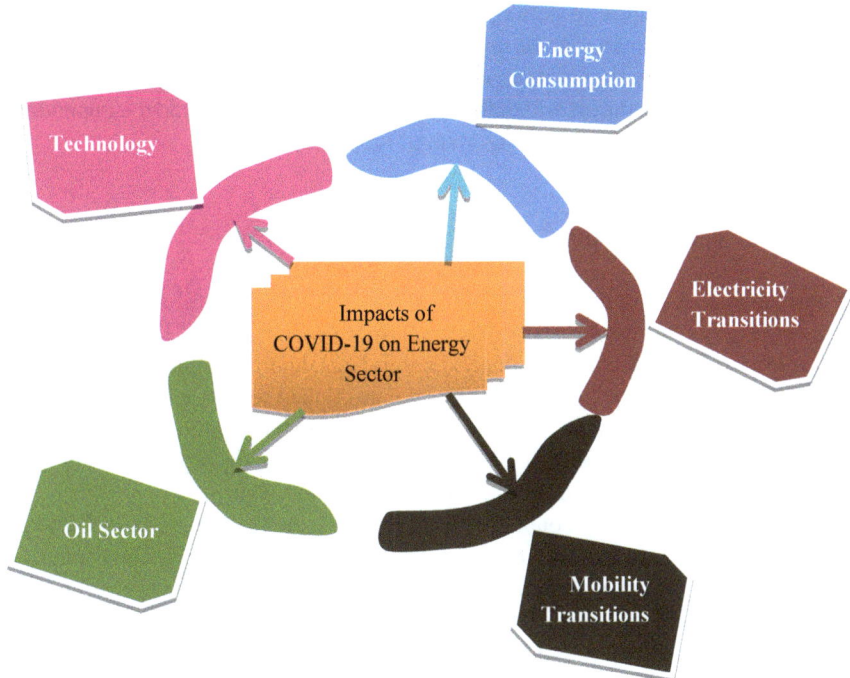

Fig. 4 COVID-19 impacts on energy sector

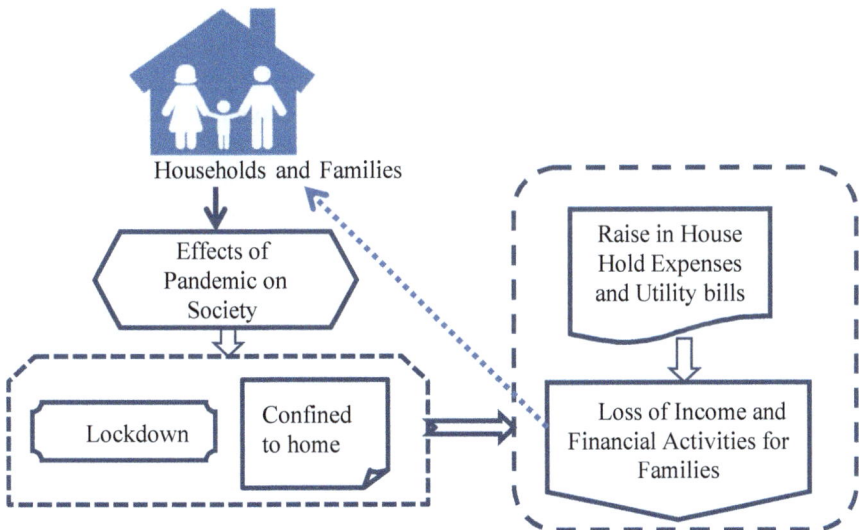

Fig. 5 Impact of COVID-19 on households

and recovery. The long-term period from mid-2020 to 2030, with the most visible implications of the decisions made now [16]. The Pandemic's effects on the nature of change in the global energy system, which has already been the subject of considerable debate [17], about the growth of the low-carbon energy policy and the impending depletion of the current fossil energy system. fuels [18].

4.1.1   The Short-Term Impacts of COVID-19 on Energy

An abnormal decline in energy demand is a more natural but still substantial one. The natural gas market's original effect was more moderate, but lack of demand has limited world production of liquefied natural gas, and the future is increasingly unlikely. The demand for natural gas and coal are linked to the production of electricity, the relative effect of the "blockade" was related to the degree and duration, and its various effects on the industrial operation, complicating the problems associated with the demand for energy, domestic gas, and electricity [19].

The authors in [20] have conducted a study on the short-term effect of COVID-19 on energy. They studied the reduction ratio by comparing energy consumption after the Pandemic from March to June 2020 with consumption before closings. By checking the effects of renewable production, seasonal trends, and weather, they predict energy use during closures.

Figure 6 shows that gasoline and jet fuel are the most significant percentage drops in energy consumption, with reductions of 30 and 50% that seems persistent in line with travel estimates in personal vehicles. Conversely, most other categories have seen minor reductions. The use of natural gas in commercial and residential buildings has decreased by approximately 20%. The general electricity demand has been decreased by less than 10%. Global electric vehicle (EV) sales are projected to decline by 43% in 20204 due to declining total car sales, combined with low gasoline prices. As with energy efficiency audits, solar installations and residential rooftop storage have plummeted. Renewable technologies have stagnated only at the utility level. Overall, renewable energy jobs fell by almost 600,000 in late April.

Fig. 6 Short-term reductions in energy consumption

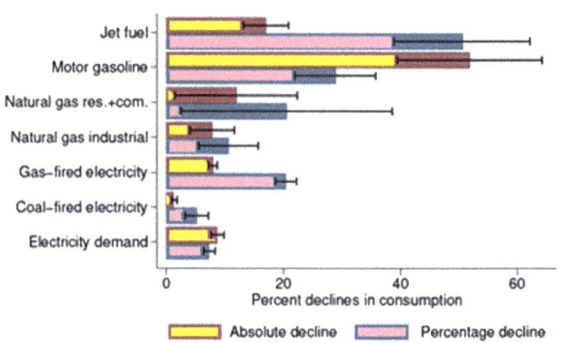

4.1.2   Long Term Impacts of COVID-19 on Energy

More than 80% of the states aren't pure exporters of fossil fuels. This transition of these fuels can be stimulated by contracting future high prices of oil and gas.Furthermore, the destruction of the demand for fossil fuels would pick up the pace, which would restrict prices. Therefore, it can be argued that investing in the low-carbon transition is a mutually beneficial policy for most states that stimulates economic restoration and reduces the price of potential fossil fuel imports [21].

With COVID-19, employers and workers can be adequately satisfied with remote work even after the threat is overcome. This could lead to some tradeoffs such as:

- It would minimize transportation but would probably expand household energy usage.
- Home energy use would increase, although commercial building use would remain virtually unchanged if office space were used equally by the remaining employees, indicating a marginal net effect.
- If teleworking is widespread, more valuable changes in commercial buildings cannot be ruled out.

The most significant long-term indirect impact of COVID-19 is expected to be on investment in the energy sector. The companies that are most likely to liquidate include startups that still have earnings to prove (Coal mining firms and firms developing low-carbon technologies). Investments in renewable energy would fall into a recession, with declining funding and poor wholesale energy rates because of reducing electricity demand [22].

This will affect investments in solar energy on roofs, electrical infrastructure, and energy conservation. It would also affect the transition to a cleaner vehicle fleet. The short-term decrease in EV sales will continue. Still, possibly most importantly, it will be difficult for cash-strapped automakers to keep investing as much in new vehicle technology to improve performance and the network of the load will be reduced. The results indicate that even rolling back a year, all renewable energy production investments would prevent deaths. However, COVID-19's energy policy response is the wild card that can change anything.

4.2   Electricity Transitions

COVID-19 Pandemic has a profound effect on the energy sector in many countries. These effects have led to a drop in demand and the prices of electricity, oil, and gas [23]. In particular, the demand for electricity in countries has decreased in recent months. This decrease results from factory detention operations. Furthermore, demand and supply trends have changed as commercial and industrial production has decreased and national demand has increased [24]. However, the medium and long-term effects of the Pandemic are difficult to predict and may be affected by the way the stimulus programs are carried out.

These stimulus packages can also promote energy changes and economic growth when they focus on the introduction of green energy technologies or improvements in energy efficiency. The Nordic energy market's implications are expected to be based on the effects of the crisis in regions of the world with the critical resources needed for smart grid energy, storage technologies, and renewable energy. Therefore, manufacturing, mining, internal stability, and trade routes would affect energy security and other industrial supplies in many areas, which had previously been vulnerable to internal security conflicts before [25].

4.3   Mobility Transitions

By influencing new sustainability niches, the COVID-19 Pandemic triggers a dramatic transition. Biogas services and low-carbon transport through electrification are promising cases of these niches in the mobility market. Sweden and Finland have developed the idea of mobility as a service (MaaS) and its applications. MaaS provides various modes of travel, accessibility packages, and payment functions [26]. Spreading these shared mobility solutions progresses is slowing down.

However, the economic effects of COVID-19 have significantly decreased the sale of electric vehicles. The wide and increasing use of electric vehicles is expected to decrease health risks compared to traditional taxis by decreasing human contact in the absence of human drivers. Therefore, COVID-19 impacts may be more favorable for the diffusion of electric vehicles (EV) [27].

4.4   Technology Sector

New advancements and innovations in cyber-physical systems and digital automation allow the introduction of decentralized production processes. These technological capabilities can provide a reduction in energy and travel resources. Also, these systems are helpful for social distancing while persevering production. "Kiva and hepatic" robots can provide the advantage that they can be directly operated over longer distances [27].

A.   **Robotics**

The development of robotic production and sourcing of flight factories offers probable environmental benefits and productivity improvements in manufacturing. Robotics can contribute to The diffusion of the building off-site with three advantages. [28]:

- Environmental: through allowing more accurate construction to bridge the gap between planned and actual energy use.

- Economical: due to increased productivity.
- Social: potentially reducing accidents on the site.

B. **Virtual Reality**

Deployment of Virtual Reality (VR) systems will reduce energy usage due to reduced transportation needs to and from work. The use of VR technology can be of great benefit in the design of the installation. Virtual reality can be used to see and explore within built environments. When infection is a primary concern in events of emergency, it may be necessary to connect these virtual systems with robots. Additionally, the 3D visualizations of Virtual facility layouts can enable robots to act in place of human operators [29].

4.5   *Oil Sector*

Oil demand is projected to decline via 20 million barrels per day in April. Also, more anomalies were observed due to a shortage of oil storage volume resulting from demand collapse [30]. In Global Review of Energy 2020, which is primarily focused on understanding the potential impacts of Covid19 [31], the latest IEA report predicts that oil demand will decline by 9%, or 9 million barrels per day, on average., reducing oil consumption until 2012. levels during the year. Compared to all other primary energy needs, this is the biggest drop (coal, gas and nuclear). There will only be growth in renewable energy (electricity). Furthermore, nothing is said about alternative renewable fuels of biological or non-biological origin or waste [32].

5   Conclusions

The coronavirus has dramatically changed human society in various ways. This paper explains the pandemic (Covid-19) impacts from a comprehensive perspective in the electricity and oil sectors. It showed how the forced shutdown has brought about rapid change and shifting oil and electricity consumption patterns. This led to a short behavioral shift in energy use. Total electricity consumption saw a marked decline after lockdown restrictions were implemented in many countries. Total electricity generation declined as demand declined, and coal generations have suffered the most. This paper also explains that oil markets have become more down during the Pandemic than in the pre-COVID-19 period. Restrictions on travel around the world cause the impact of COVID-19 on crude oil prices. This is essential to many industries, including transportation, oil exploration, hospitality, and production companies. The negative impact of the epidemic on the efficiency of the commodity market is evident. The percentage of renewable generations has grown, although rates of contraction are also become higher. The policy tools applied to shape, direct, and manage transitions in energy systems must be reassessed. These policies can lead to a strong energy shift in consumption patterns. The severity of COVID-19 and its long-term impact on the electric power and Oil industry remains to be seen as an important area for future research. There is a great need to include measures to enhance systems' resilience to respond to crises and consider the lessons learned they provide to achieve changes in energy systems.

References

1. World Health Organization (WHO): WHO director-general's opening remarks at the media briefing on COVID-19. https://www.who.int/dg/speeches/detail/who-director-general-sopening-remarks-at-the-media-briefing-on-COVID-19-11-march-2020 (2020). Accessed 5 Aug 2020
2. Cohen, J.: Scientists' strongly condemn' rumours and conspiracy theories about the origin of the coronavirus outbreak. Science (2020)
3. Smalley, R.E.: Our energy challenge. a lecture at the james A. baker III institute for public policy of rice University. https://www.bakerinstitute.org/media/files/Research/3cf8b6ec/NANO_S malleylecture.pdf (2005). Accessed 12 May 2020
4. Oil and Coronavirus Shocks Add Pressure for MEA Sovereigns. https://www.fitchratings.com/research/sovereigns/oil-coronavirus-shocks-add-pressure-formeasovereigns-10-03-2020 (2020). Accessed 10 Aug 2020
5. Fulwood, M.: $2 gas in Europe is here: who will blink first?. https://www.oxfordenergy.org/publications/2-gas-in-europe-is-here-who-will-blink-first/?v=7516fd43adaa (2020). Accessed 10 Aug 2020
6. Wagh, M.M., Kulkarni, V.V.: Modeling and optimization of integration of Renewable Energy Resources (RER) for minimum energy cost, minimum CO2 emissions and sustainable development, in recent years: a review. Mater. Today. Proc. 5(1), 11–21 (2018)
7. Qarnain, S.S., Muthuvel, S., Bathrinath, S.: Analyzing factors necessitating conservation of energy in residential buildings of Indian subcontinent: a dematel approach. Mater. Today. Proc. (2020)
8. Abu-Rayash, A., Dincer, I.: Analysis of mobility trends during the COVID-19 coronavirus pandemic: exploring the impacts on global aviation and travel in selected cities. Energy Res. Social Sci. 68, (2020)
9. Bahmanyar, A., Estebsari, A., Ernst, D.: The impact of different COVID-19 containment measures on electricity consumption in Europe. Energy Res. Social Sci. 68, (2020)
10. Jefferson, M.: A crude future? COVID-19 s challenges for oil demand, supply and prices. Energy Res. Social Sci. 68, (2020)
11. Jiang, P., Van Fan, Y., Klemeš, J.J.: Impacts of COVID-19 on energy demand and consumption: challenges, lessons and emerging opportunities. Appl. Energy 116441 (2021)
12. Crude oil facts. https://www.nrcan.gc.ca/science-data/data-analysis/energy-data-analysis/energy-facts/crude-oil-facts/20064,lastvisited14/2/2021
13. US Field Production of Crude Oil. https://www.eia.gov/dnav/pet/hist/LeafHandler.ashx?n=PET&s=MCRFPUS2&f=M,lastvisited14/2/2021
14. Fu, M., Shen, H.: COVID-19 and corporate performance in the energy industry. Energy Res. Lett. 1(1), 12967 (2020)
15. The Earth Institute, Columbia University.: New Data Suggest COVID-19 Is Shifting the Burden of Energy Costs to Households. https://blogs.ei.columbia.edu/2020/04/21/COVID-19-energy-costs-households/ (2020, April 21). Retrieved 24 April 2020
16. Steffen, B., Egli, F., Pahle, M., Schmidt, T.S.: Navigating the clean energy transition in the COVID-19 Crisis. Joule (2020). https://doi.org/10.1016/j.joule.2020.04.011 (2020, April 21). Accessed 3 Aug 2020

17. Sovacool, B.: The history and politics of energy transitions: comparing contested views and finding common ground. In: Arent, D., Arndt, C., Miller, M., Tarp, F., Zinaman, O., (eds.) The Political Economy of Clean Energy Transitions. Oxford: Oxford Scholarship. https://doi.org/10.1093/oso/9780198802242.003.0002

18. World Economic Forum: The speed of the energy transition: gradual or rapid? geneva: WEF. http://www3.weforum.org/docs/WEF_the_speed_of_the_energy_transition.pdf (2019). Accessed 3 Aug 2020

19. Honoré, A.: Natural gas demand in europe: the impacts of COVOD-19 and other influences in 2020. Oxford: Oxford Institute for Energy Studies, Oxford Energy Comment (2020). Accessed 17 Aug 2020

20. Gillingham, K.T., Knittel, C.R., Li, J., Ovaere, M., Reguant, M.: The short-run and long-run effects of COVID-19 on energy and the environment. Joule (2020)

21. Okoh, A.S.: Oil Mortality in Post-Fossil Fuel Era Nigeria: Beyond the Oil Age. Springer Nature (2020)

22. Hepburn, C., O'Callaghan, B., Stern, N., Stiglitz, J., Zenghelis, D.: 'Will COVID-19 fiscal recovery packages accelerate or retard progress on climate change?', Oxford Smith School of Enterprise and the Environment, Working Paper 20–02 (2020)

23. Mylenka, T., Novyk, B.: Impact of COVID-19 on the global energy sector. https://www.pv-magazine.com/2020/04/24/impact-of-COVID-19-on-the-global-energysector/ (2020). Accessed 3 Aug 2020

24. Lempriere, M.: COVID-19 and flexibility: pandemic to change established and expected patterns, Current News. https://www.current-news.co.uk/news/COVID-19-and-flexibility-pandemic-to-change-established-and-expected-patterns (2020). Accessed 27 July 2020

25. NordPool, Market data. https://www.nordpoolgroup.com/Market-data1/Powersystemdata/Production1/Production1/FI/Hourly4/?view=table (2020). Accessed 5 May 2020

26. Government decree blocks providers from cutting utilities for three months: Buenos Aires Times. Retrieved https://www.batimes.com.ar (2020, July25)

27. New Rules Affecting Foreigners in Germany as of April 2020. https://www.germany-visa.org/news/ninenew-rules-affecting-foreigners-in-germany-as-of-april-2020/ (2020, April 9). Retrieved 11 July 2020

28. A brave new world: Lessons from the COVID-19 Pandemic for transitioning to sustainable supply and production

29. Iuorio, O., Wallace, A., Simpson, K.: Prefabs in the North of England: technological, environmental, and social innovations. Sustainability **11**(14), 3884 (2019)

30. Corkery, M., Gelles, D.: Robots welcome to take over, as pandemic accelerates automation. The New York Times April 10. https://www.nytimes.com/2020/04/10/business/coronavirus-workplace-automation.html. Accessed 27 April 2020

31. Reed, S.: The world is running out of places to store its oil. The New York Times, March 27 2020. https://www.nytimes.com/2020/03/26/business/energy-environment/oil-storage.html (2020). Accessed 15 April 2020

32. IEA: Global Energy Review 2020–The impacts of COVID-19 crisis on global energy demand and CO2 emissions (2020)

COVID-19 Outbreak and Its Effect on Global Environment Sustainable System: Recommendation and Future Challenges

Amira S. Mahmoud, Mahmoud Y. Shams ⓘ, and Aboul Ella Hassanien ⓘ

Abstract Long-standing exposure to concentrations of air pollutants is known to cause the inflammation of the lungs chronic, a state that might enable the increased severity of new coronavirus-induced COVID-19 pandemic called (SARS-CoV-2), which is the main cause of the epidemic recently confirmed by the World Health Organization (WHO). In recent times, the infectious disease caused by the new coronavirus began in China. COVID-19 has rapidly spread worldwide, presenting the entire human population with immense health, economic, environmental, and social challenges. Owing to the pandemic of COVID-19, Since mid-March 2020, human activities have been increasingly limited in many nations, and this is a radical experiment to show the efficacy of restricted pollution. Air contaminants' Impact as delicate (PM2.5) particulate matter, NO_2, SO_2, CO, and excess fatalities within the half-moon of 2020 are studied in this chapter. Besides, we present a future planning strategy to fight the COVID-19 pandemic resulted from pollution using Artificial Intelligence (AI) tools. Furthermore, in this chapter, we study the influence of the immediate organized processes realized by Egypt's government due to the recent COVID-19 pandemic with a straight affect on the air quality improvement.

Keywords Environmental COVID-19 effect · Air pollution · Artificial intelligence

A. S. Mahmoud
Department of Environmental Studies, Institute of Graduate Studies and Research, Alexandria University, 163 Horreya Avenue, El-Shatby, P.O. Box 832, Alexandria, Egypt

M. Y. Shams (✉)
Department of Machine Learning and Information Retrieval, Faculty of Artificial Intelligence, Kafrelsheikh University, Kafr El Sheikh 33511, Egypt
e-mail: mahmoud.yasin@ai.kfs.edu.eg

A. E. Hassanien
Faculty of Computer and Artificial Intelligence, Cairo University, Cairo, Egypt
e-mail: aboitcairo@cu.edu.eg

A. S. Mahmoud · M. Y. Shams · A. E. Hassanien
Scientific Research Group in Egypt (SRGE), Cairo, Egypt

1   Introduction

The first identification of coronavirus disease was in December 2019 in Wuhan, China [1, 2]. The SARS-CoV-2 or COVID-19 disease later spread to other countries in Asia, Africa, and Europe. More especially in Europe, Italy, Spain, France, and United Kingdom are affected directly to down break results from COVID-19 pandemic. Health Organization declared that COVID-19 an epidemic on March 11, 2020, after the confirmed cases were spread, reaching 118,319 cases in over 110 countries. For example, on June 4, 2020, the virus has spread further to succeed in almost every country globally. Nearly 220, consistent with things provided by the WHO COVID-19 Dashboard on December 26, 2020, The overall number of confirmed cases of COVID-19 surpassed the number of 79,000,000, while the deaths cases exceeded 1,744,235 persons. The effect of COVID-19 spread on the environment is often classified generally into two categories. First is the positive effect within the surrounding environment, while the second is the negative COVID-19 spread effect on the environment.

Figure 1 briefly illustrates the Impact of COVID-19 on the environment. In general, we have observed that improving the level of air pollution as well as reducing environmental noise is achieved. On the other hand, the main negative Impact of COVID-19 on the environment is the medical waste generated by masks and gloves.

To trace the variability and changes of the recent pandemic, Fig. 2 investigates the globally confirmed cases of COVID-19 changes reported by WHO. In contrast, Fig. 3 shows the corresponding situation by Country, Territory, and Area.

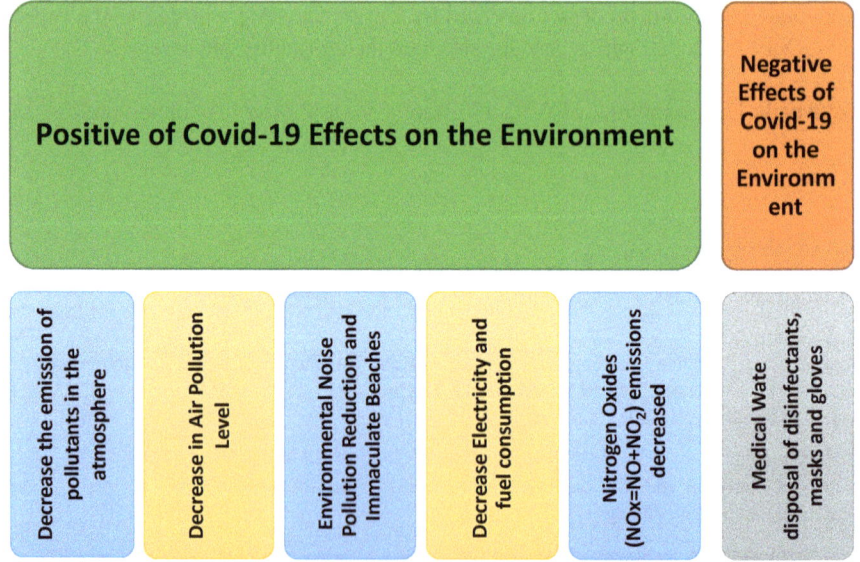

Fig. 1   The Negative and positive effects of COVID-19 on the environment

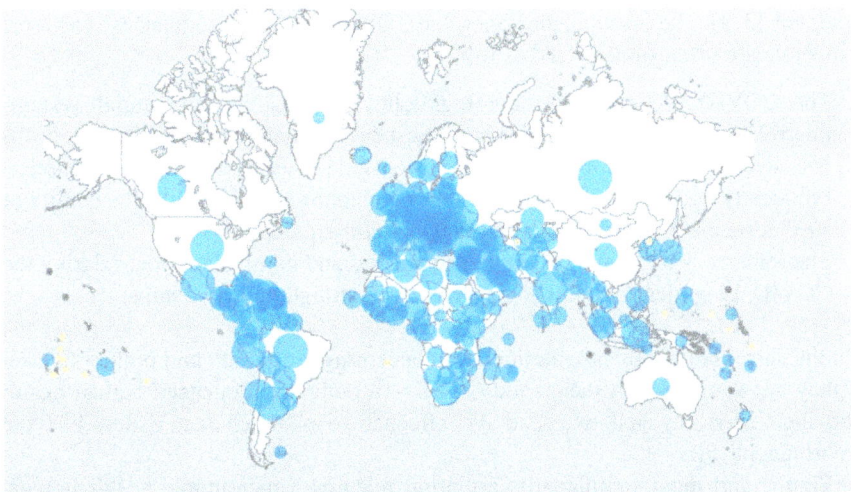

Fig. 2 Globally, at 5:03 pm CET, 26/12/2020, the confirmed cases of COVID-19 were 78,604,532, including 1,744,235 deaths, as reported to WHO

Name	Cases - cumulative total ⇄↓	Cases - newly reported in last 24 hours	Deaths - cumulative total	Deaths - newly reported in last 24 hours	Transmission Classification
Global	78,604,532	401,176	1,744,235	7,335	
United States of …	18,311,405	0	323,527	0	Community transmission
India	10,169,118	22,273	147,343	251	Clusters of cases
Brazil	7,425,593	60,076	190,006	786	Community transmission
Russian Federati…	3,021,964	,29,258	54,226	567	Clusters of cases
France	2,505,074	20,199	62,051	159	Community transmission
The United King…	2,221,316	32,725	70,195	570	Community transmission
Italy	2,028,354	19,037	71,359	459	Clusters of cases
Spain	1,854,951	0	49,824	0	Community transmission
Germany	1,627,103	14,455	29,422	240	Clusters of cases
Argentina	1,571,680	7,815	42,392	78	Community transmission
Colombia	1,559,766	14,940	41,454	281	Community transmission
Mexico	1,362,584	12,485	121,172	861	Community transmission

Fig. 3 Situation by Country, Territory & Area WHO Coronavirus Disease (COVID-19) Dashboard (https://covid19.who.int/table)

The COVID-19 epidemic had a significant influence on personal lives around the world. It emphasizes what we all distinguish the COVID-19 short-term effects of on the surrounding environment. It considers what is often learned from these influences and how they will help shape future decision-making. As reported by the Environment Agency, COVID-19 and Europe's environment influence a worldwide epidemic [3, 4]. The essential messages that COVID-19 has consequences of a global pandemic are often summarized as follows:

- The COVID-19 epidemic more highlights the societal and natural systems interrelations: societal depends on a robust environmental network,
- Bio-diversity loss and intensive food systems make the diseases more zoonotic,
- Frequently associated with social-environmental in-equalities, like the quality of air that seems to be affected by COVID-19 consequences,
- The reliance is increased on single-use plastics and nominal oil prices during the COVID-19 epidemic. It has some short-term, straight, positive influences.

On the surrounding climate, particularly concerning air quality and pollution, even if they are temporary. It means the COVID-19 outbreak influenced human health and their economy and extended its influence to other characteristics, like the environment [5].

Firstly, and most significantly, pollution features a measurable health impact. Globally, pollution is that the second leading explanation for death from non-communicable diseases after tobacco smoking. The air pollution in the main public health regions are mainly included (PM), tropospheric (ground-level) ozone (O_3), dioxide (NO_2), and sulfur dioxide (SO_2), which may affect many organs and systems. The evidence is most vital for cardiovascular and respiratory effects. Other possible health outcomes include metabolic effects, atherosclerosis, impaired neurological and lung development in children and even an association with neurodegenerative diseases. It's also a drag of inequity, as pollution particularly affects those already disadvantaged or vulnerable: people cannot choose the air they breathe. Secondly, some air pollutants—particularly black carbon (a component of PM) and tropospheric O_3 – arc also short-lived climate pollutants linked with health effects and near-term warming of the earth.

Air pollution, especially in developing nations like Egypt, has emerged as a growing concern worldwide. The study found that 19,200 persons died early and more than 3 billion days lived with the disease in Egypt in 2017 due to urban PM2.5 air pollution especially, in Greater Cairo Egypt, regarding the World Bank report, in 2019, the environmental cost related to degradation: water and air pollution [6].

Furthermore, recent studies suggest that high air pollution levels significantly prolonged exposure increase the mortality and vulnerability rates as a result of COVID-19 [7, 8]. More precisely, The relative role of air pollution and surrounding aerosols in the distribution of air pollution, viruses and mortality rates remains debated by the global and international scientific community [9].

Nevertheless, there is some evidence that, although not ultimately confirmed, SARS-CoV-2 can be spread by aerosols, so the surface of sanitization, good ventilation, and clean environments help to restrict the spread of the virus [8, 9].

The Indian cities have always been among the highest 20 most polluted cities globally and have met the planet Health Organization. Therefore the Central Pollution Control Panel (CPCP) recommended environmental air quality standards. A study of air quality modeling in India is presented by Garaga et al. [10], by which they introduce the proper formulation of air quality abatement strategies. In comparison, Guo et al. [11] proposed local and regional sources' influences on PM2. 5 and its general impacts in north India. The National Clean Air Program (NCAP) In 2019 unveiled a five-year action decide to reduce PM by 30% nationally to manage the country's extreme pollution. In the references cited in [12–14], the most promising solutions presented by a well-organized implementation are presented to immune the percentages of resulting air expected emissions.

More exactly, we can notice that atmospheric processes that assess absorptions of pollutants in the air that are non-linear and altering the meteorology play a vital role in pollution generation. Chinese five-year clean air is one of the most known examples by which the action plan has been performed to enhance China's air quality significantly, as presented by Li et al. [15]. Nevertheless, because of unfavorable meteorology, the height concentrations of PM2.5 during winter episodes didn't decrease. Furthermore, as presented in Wang et al. [16] and Zhang et al. [17] a conduction of nearly thirty-three percent reduction in nitrate by regulating pollution compensated using meteorology in the eastern US. A simulation wiped out by China showed that metrology played a critical role in emission creation, which was not prevented in January and February 2020 during the lockdown [16].

Mohamed et al. [5] address the effect of the COVID-19 pandemic on air pollution levels in Egypt by researching some environmental parameters as Aerosol Absorption Particles (AAP), carbon monoxide (CO), greenhouse gases (GHG), nitrogen dioxide (NO_2), and ozone (O_3), noise and solid waste from municipal and medical sources.

- The data used in their analysis was gathered from satellite data sets for air-pollution levels from the National Aeronautics and Space Administration (NASA) and the European Space Agency.
- Data for the 2020 period were associated to the related months for the specified base-line period (2015-2019), focusing on Cairo and Alexandria's governorates to determine the closure effect on air pollution levels in Egypt.

Their paper introduces the tale of Egypt's COVID-19 from multiple perspectives, including the evolution of reported cases of COVID-19, government containment initiatives, the effects on the national energy, and country's economical use, to assess the Impact on other environmental indicators accurately and the levels of air pollution studied. For additional ecological parameters, a critical relationship has been observed between locking down COVID-19 and reducing ecological noise, beach, soil, and groundwater pollution. As for noise, this was supported by official government announcements claiming that the amount of environmental noise in Egypt had decreased by about 75% during the closure period. On the one hand, there are some negative effects, including increased medical, solid waste (70–300 tons/day), urban solid waste, and a less effective recycling process for solid waste. The AAI index in Cairo and Alexandria governorates decreased by about 30%; nitrogen dioxide decreased by 15 and 33%. In comparison, in the provinces of Cairo and Alexandria, ozone levels rose by around 2%.

It is possible to assume that It can be concluded that the containment measures adopted during the COVID-19 pandemic had a positive and negative climate effect. The Positive Effects of the environment are not sustainable, as, before the pandemic, post-lockdown degradation is expected. Therefore, to protect the climate in Egypt, stricter laws have to be enacted [5]. Within the first four months, the spread of Coronavirus disease 2019 (COVID 19), which was initially detected in Wuhan, China, resulted in more than one million cases worldwide. In many nations around the world, it has led to a lockout. A first foreign travel advisory has been released, setting restrictions on travel to China, the Republic of Korea, Iran, Italy, and Japan. March 11 following a sudden spike in COVID-1919 in the region (https://www.mohfw.gov.in/), while India's first verified case was on January 30, 2020.

2 Case Studies in Air Pollution

Air pollution usually occurs in three cases, and the first was the Gaseous that directly released (CO, SO_2, NO, H_2S, CH_2O) molecules or formed from other contaminants (NO_2, O_3) in the atmosphere. Liquid: (benzene) C_6H_6, the fog droplets are covering other pollutants. Solid: numerous particles airborne, including smoke, combustion materials, and pollen, small enough to be inhaled. Gases and liquids have a specific chemical that enables the control of the environment, but in certain substances, there are, Particulate Matter (PM). The Soot denotes the black carbon (BC) or Elemental Carbon (EC), but size ranges generally govern airborne particles without chemical composition [18–20].

2.1 Descriptions of Particles

There is a long tradition of worries about airborne particles, from smoke to grit. The first paper discussed the air pollution were created in 1913 to distinguish between air pollution and the related health associated the concentrations of Soot composed on plates horizontally at various locations. Perhaps their correlation with pneumonia and tuberculosis mortality rates. For decades, Soot has been subject to pollution in major countries, especially the United States, where training has shown robust connections between smoke and death's daily levels. For airborne particle classification, the scale is the primary metric based on aerodynamic diameter, which considers density mutually and microns (μm) [21].

The Total Suspended Particulate (TSP) is less than 50 μm, and less than ten μm in the corresponding inhalable PM, besides <2.5 μm "fine" PM. The Ultra-Fine Particles (UFP) are lesser than about 0.1 μm. They contain many reduced Nano-particles representing the atmospheric amounts, which are determined as particle numbers in its place of mass. Since the particle mass upgrades with the cube diameter, a PM-10 particle would evaluate as 64-times-quite as in PM2.5 particle and corresponding a billion UFPs. Therefore, given the air's monitored contaminants, the atmosphere is composed by observing the pollution system as just a print of what we breathe [22].

The United States began a national TSP air sampling program in the 1950s that involved drawing ambient air through a filter and measuring the collected tissue. The initial impulse was the search for radioactive dust from previous nuclear weapons tests; The network also collected data on a wide variety of inorganic and elemental particulate components. The Soot was initially evaluated in the filter sediments by the degree of darkening and now by qualitative analysis. In 1979, the Environmental Protection Agency (EPA) commissioned a nationwide network of selective fine-sized particle samplers to collect fine, coarse, and inhalable particles. This data was not incorporated into the EPA regulatory database and is generally ignored by epidemiologists, but was subsequently used in various human health studies. Studying the health effects of PM2.5 began in 1981, citing concerns that these particles might damage the lower lung and surrounding tissues. The EPA started the regular nationwide sampling of atmospheric concentrations in 1999. The contrast between "inhalation" comes from respiratory dynamics studies that deposit larger particles in the nose and throat, slightly smaller particles in the upper respiratory tract, and PM2.5 in the lower respiratory tract. UFPs can easily deposit and travel into the bloodstream in the alveoli. However, inhaled particles' actual behavior is considerably more complex and requires both sedimentation and filtering processes. Samples of lung tissue dissected from Mexico City showed accumulated carbonate particles. The masses of elementary carbon (EC) are rarely seen in "clean" cities of the lungs. Most of the particles were smaller than 0.25 μm, soluble particles such as PM sulfate content were not visited, and the smaller particles were challenging to enumerate [23–25].

2.2 Sources and Distributions of Ambient PM2.5

The characteristic of particle size varies from numerous emission sources and normal measurement ranges; medians are 2.5 and 10 μm, not maximum aerodynamic diameter values.

All three smallest size ranges, PM2.5, ultrafine, and PM-10, provide various natural and chemical emission and composition resources. For example, a PM2.5 of collected samples are in a huge northeastern city subject to fly ash, oil smoke, and H_2SO_4 mist, whereas different types of dust could dominate one from the southwest. Indoor models of PM2.5 are likely to contain ambient cigarette smoke, black (elemental) gas, and various organic compounds. As the Clean Air Act mandates, this variety makes it challenging to formulate good national pollution monitoring strategies [26].

In the atmosphere, PM dispersion and interactions often depend on particle size and composition. As predictors of health effects, The resulting variability complicates their usefulness. Besides inorganic salts such as $(NH_4)_2SO_4$ and NH_4NO_3, PM samples can include the periodic table elements. These combinations are interested in particular. Meanwhile, they are regulated shaped in the atmosphere from emissions of SO_2 and NO_2. These secondary contaminants are less thoroughly controlled and make epidemiological cause and effect less reliable; more information can be given by mathematical modeling [27–29].

Hence, Particulate Matter (PM)'s role in the spreading and increase in the morbidity and mortality of COVID-19 are studied. Firstly, the correlation between PM and COVID-19 is the basis for starting a broader study. This relationship between the variables is positive, but their function must be understood, as all the reviews mentioned indicate a rise in the global incidence of the Covid-19 virus. Long-term and short-term exposure to high levels of toxins is correlated with it. It may be of interest to conduct systematic research, position particle collection units at strategic points around the world, such as hospital departments with a low ventilation rate, and examine microorganisms present in the hunt for COVID-19 viral metagenomics [30, 31].

This hypothesis will be supported by preliminary data obtained by Setti et al. [32]. Since Sars-Cov-2 RNA could be illustrated in 34 external collected samples on three Weeks, between February 21 and March 13, at the industrial site of Bergamo County, on the other hand, there are chronic inflammatory conditions in the population who live in areas with a high proportion of pollutants (taking into account gender, age and genetic factors), which makes them more vulnerable to respiratory diseases [33].

3 The Impacts on the Atmospheric Environment Caused by COVID-19

3.1 Southeast Asia Region

Malaysia and other Southeast Asian (SEA) nations have also introduced shutdowns to various degrees to avoid the disease's spread, which has benefited the natural environment. In addition to assessing the impact on air pollution and economic development, and evaluation study is presented by Kanniah et al. [34] to reduce anthropogenic emissions due to COVID-19 and the related government initiatives to limit its growth. They used Himawari-8 satellite Aerosol Optical Depth (AOD) observations to quantify the changes in aerosol and air pollutants associated with the general shutdown of anthropogenic pollutants And the column density of Aura-OMI tropospheric NO_2 over SEA and ground-based emission measurements at various stations across Malaysia. The lockout has resulted in a major decrease in AOD over the SEA and emission outflows around the oceanic regions. In areas not affected by seasonal biomass burning, a substantial reduction in tropospheric NO_2 (27–30%) was observed, on the other hand, in the decrease in AOD over the SEA and the outflow of pollution over the ocean areas.

In Malaysia, in general, the concentrations of PM10, PM2.5, NO_2, SO_2, and CO decreased by 26–31%, 23–32%, 63–64%, 9–20%, and 25–31%, respectively, in urban areas during the shutdown process, compared to the same periods in 2018 and 2019. In the commercial, suburban and rural locations of the country, notable declines are also seen. The elimination of negligible and health-damaging air pollution must be calculated by climate change studies and health-related research and air quality.

3.2 Northern Italy Region

The research empirically investigates the ecological association between long-term fine particulate matter (PM2.5) concentrations in the area and excess fatalities in Northern Italy's municipalities in the first quarter of 2020. The research illustrates potential urbanization-related regional contributing effects. It may have affected the spread of SARS-CoV-2 and the related mortality of COVID-19. To determine the typical spatial distribution of ambient PM2.5 concentration and ambient PM2.5 concentration,Our epidemiological study uses geographic knowledge (e.g., municipalities) and negative binomial regression for excess mortality. The positive association of ambient PM2.5 levels with excess mortality associated with the COVID-19 epidemic in Northern Italy has been shown in our research. The estimates indicate that a one-unit rise in PM2.5 concentration ($\mu g/m^3$) is correlated with a 95% increase in mortality associated with COVID-19 [35].

3.3 United Arab Emirates Region

During the recent COVID-19 pandemic, extreme protective lockout measures have been implemented by the Northern Emirates of the United Arab Emirates (NE-UAE). A lockout was introduced on April 1, 2020. Because of lower emissions, it was believed that air quality and Surface Urban Heat Island Intensity (SUHII) had been substantially improved. In this analysis, NEUAE tested three Nitrogen Dioxide parameters (NO_2), Aerosol Optical Depth (AOD), and SUHII variables.

The core results showed that the average NO_2, AOD, and SUHII levels decreased by 23.7%, 3.7%, and 19.2%, respectively. During lockdown times, compared to the same time in 2019. Validation of outcomes indicates a high agreement between the values expected and calculated. The agreement for NO_2, AOD, and night LST, respectively, was as high as $R2 = 0.7$, $R2 = 0.6$, and $R2 = 0.68$, suggesting substantial positive linear correlations. The present study concludes that the lockdown measures have significantly decreased NO_2, AOD, and SUHII due to decreasing automotive and industrial emissions in the NE-UAE. Furthermore, as they are mostly related to dust's normal occurrence, the aerosols did not change significantly. The current investigation found that NO_2, AOD, and SUHII concentrations decreased across the NE-UAE during the pandemic lockout. Compared to the same time, the highest average decrease in NO_2 (23.7%) was followed by SUHII (19.2%) and AOD (3.7%) over the lockdown period [36].

4   Effects of COVID-19 on Air Quality in Egypt

Because of the geographical position of Egypt, many factors affected the pollution in Egypt. These factors include industrialization and change in overpopulation, as well as geomorphological features. Moreover, the high wind speeds up the flow of dust from the desert during the spring season and leads to enlarged pollution levels all over the nation [37–39]. Cairo and Nile-Delta regions are the most heavily populated and highly polluted regions in Egypt.

Data surveys from January to June in 2018, 2019, and 2020 showed variability in the three-year emissions and concentrations. After applying the national lockdown in Egypt on March 15, 2020, a decrease in the concentration of NO_2, CO, O_3, and aerosol optical depth AOD compared to the previous 2 years has been observed. The partial lockdown applied during the daytime, and the full lockdown could return to this reduction. The CO distribution and variability study does not indicate dramatic changes following the lockdown implementation. The CO pattern began to decrease every year from May due to changes in meteorological and climatic conditions. This time coincided with the spring season, which has unpredictable conditions with a slightly higher temperature and wind speed than the winter season (January-February). This time of the year shows a general usual growing pattern as a transitional period from spring to summer, along with days of heatwaves and heat islands [40]. Further, in 2020, atmospheric ozone analysis showed slightly lower values than in 2018 and 2019, especially after introducing control measures due to the spread of COVID-19.

The ozone curve and aerosols levels show a higher decrease in 2020 than in previous years, which in return is the function of the lockdown, which represents the decrease in concentration and spatial distribution and low emissions [40].

We can see the decrease in the Nile Delta locale and Greater Cairo, the exceptionally populated metropolitan district in Egypt. This district has colossal day-by-day discharges of poisons because of mechanical exercises and transportation, and other foundation contamination sources. During the extended vacation time (total lockdown), it is seen that family unit and metropolitan outflows are diminished because a huge populace from Greater Cairo accepted this open door and moved to their external cause urban communities and country zones outside Cairo. This brought about a decrease of social and financial exercises nearby and therefore diminished toxins' outflows. Additionally, the spatial contrasts in CO fixation dispersion are seen in various locales covering North Egypt. Everywhere in the world, ozone has shown high spatial varieties. Changes are found in the Nile delta and Western desert for vaporized filling. Among 2020 and the past two years, there were normal varieties in the investigated follow gases and vaporized filling focuses. There has been a no more noteworthy improvement in a few days, and frequently the qualities are higher in 2020, which could get back to the incompletely presented lockout during the daytime. It shows that air quality has improved the country over because of the decrease of poison discharges. NO_2 consequently diminished by 45.5%, CO discharges diminished by 46.23%, ozone focus diminished.

In conclusion, for NO_2, CO, O_3 and AOD, the contaminants' concentrations decreased significantly by 45.5%, 46.23%, 61.1%, and 68.5%, respectively. That confirms the 46% reduction in NO_2 concentration and spatial distribution. The lesson learned from this pandemic of COVID-19 is to find an alternative measure of mitigation to enhance the air quality for human health. The lockdown is expected to be a successful alternative mitigation method to reduce air pollution and improve air quality. However, we realize that this form of mitigation is not applicable because it has social and economic impacts; it affects its entire economy and the population's income [41].

5   Artificial Intelligence to Fight the Spread of COVID-19 in the Presence of Air Pollution

Artificial intelligence (AI) has recently become a practical method to combat the coronavirus's spread [42–46]. There is an excellent effect in the fact of air pollution results from various sensors. In this chapter, we study the status of each patient infected by COVID-19, which undergoes more and more severe infection due to three critical factors; air pollution, home quarantine, and finally crowded and non-crowded regions.

By tracking and observing the mentioned situations, especially in the home quarantine period, we noticed a significant improvement in air pollution's surrounding environment. Therefore, we plan to use statistical analysis to predict the patients' COVID-19 status in air pollution. Figure 4 demonstrates the planning strategy to

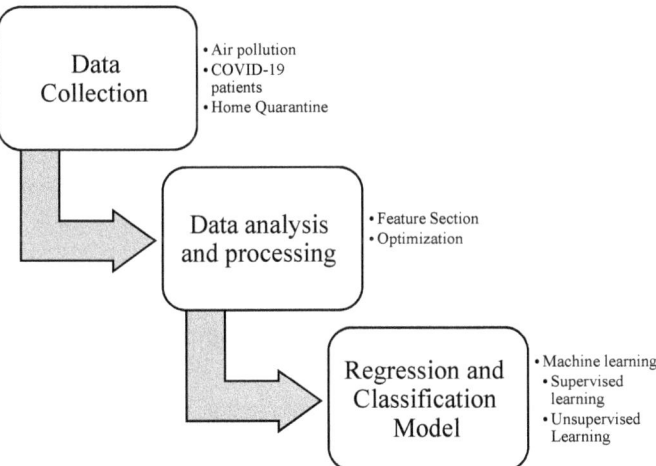

Fig. 4 The general structure of the suggested AI classification and regression model based on ML approaches

combat the development of COVID-19 in the presence of air pollution. During the COVID-19 pandemic, the data can be collected and arranged to track the environmental status variability during home quarantine and/or crowded regions to study the air pollution effect given the previously mentioned parameters independent variables. However, given COVID-19 patients in the presence of the data above and air pollution, we can classify, predict, and track these independent parameters' Impact in determining patients' cases.

AI's major challenges are the dataset collections containing the following attributes; air pollution data, home quarantine periods, crowded and non-crowded regions, and patients cases. We seek to collect this data recently to allow AI to build architecture or model based on Machine Learning (ML)approaches able to classify, predict and track the future variability of the surrounding environment and the whole effect during the COVID-19 pandemic. The general schematic diagram of the suggested AI model based on ML is shown in Fig. 4.

6   Linkages Between COVID-19 Patients and Air Pollution

Air pollution is a significant risk factor for both severe and persistent respiratory and cardiovascular diseases. People with these medical conditions are thought to be at a greater risk of developing severe COVID-19 infection disease; Thus, air pollution is likely to be a factor contributing to the health burden caused by COVID-19. However, during the global COVID-19 pandemic, we also saw a significant, albeit short-lived, air pollution reduction across cities. This reduction is most pronounced in nitrogen oxides (NO_x), a pollutant closely related to traffic and one of the activities most affected by lockdown measures. European data for some cities showed a drop of nearly 50%, and in some cases up to 70%, in NO_2 levels compared to pre-lockdown values.

Coronavirus, COVID-19, is an unfolding tragedy. Still, at the same time, it has given us an unprecedented opportunity to witness how policies related to transportation, the way people work, study and consume can be taken advantage of as we collectively move towards achieving a "new normal" that can provide environmental and health benefits.

7   Recommendation and Future Challenges

This chapter highlights air pollution and environmental factors in COVID-19 patients and how all these factors and parameters affected human health and predict the patient's cases during a recent pandemic. Moreover, meeting health-based standards for common air pollutants are the major demands for building healthcare system depends upon studying the effect of air pollution on human health and limiting climate change by looking at the global climate that the associated environmental system consists of earth, Sun, oceans, wind, rain snow, forests, deserts and savannas, and everything people do, too. The environment of a place, for example, Cairo, can be described as its mild climate, varying temperatures during the year, etc. Furthermore, we need to reduce the risks from air pollutants and protect the stratospheric ozone layer against degradation.

We propose a suggestion activity for both public or potentially worldwide a total lockdown for 2–3 days each 2–3 months that will upgrade our air quality and improve the planet's wellbeing climate. The headway of satellite sensors and the short return to time would empower to screen the effect of a particular proposed lookdown.

In this chapter, AI can overcome the future challenges on the environment and, therefore, enhance the prediction system for discovering and diagnosing patients during this pandemic. Thus, the decision-maker and digenesis can perform the evaluation based on the classification and regression techniques using AI tools. Hence, the presence of Internet of Thing (IoT) sensors will provide the data for us to be used and analyzed using AI approaches.

Using deep learning for extensive data analysis and 4 and 5G using blockchain methodologies can help us collect the data and handle it more manageable and track and predict the selected features or parameters used to build a successful prediction model.

References

1. Huang, C., Wang, Y., Li, X., Ren, L., Zhao, J., Hu, Y., Zhang, L., et al.: Clinical features of patients infected with 2019 novel coronavirus in Wuhan, China. Lancet **395**(10223), 497–506 (2020)
2. Banerjee, D.: The COVID-19 outbreak: crucial role the psychiatrists can play. Asian J. Psychiatry **50**, (2020)
3. Ordóñez, C., Garrido-Perez, J.M., García-Herrera, R.: Early spring near-surface ozone in Europe during the COVID-19 shutdown: meteorological effects outweigh emission changes. Sci. Total Environ. **747**, (2020)
4. Menut, L., Bessagnet, B., Siour, G., Mailler, S., Pennel, R., Cholakian, A.: Impact of lockdown measures to combat Covid-19 on air quality over western Europe. Sci. Total Environ. **741**, (2020)
5. Mostafa, M.K., Gamal, G., Wafiq, A.: The impact of COVID 19 on air pollution levels and other environmental indicators-A case study of Egypt. J. Environ. Manag. **277**, 111496 (2020)
6. Larsen, B.: Arab Republic of Egypt-Cost of Environmental Degradation: Air and Water Pollution. World Bank (2019)
7. Contini, D., Costabile, F.: Does air pollution influence COVID-19 outbreaks? Atmosphere **11**(4), 377 (2020). https://doi.org/10.3390/atmos11040377
8. Liu, Y., Ning, Z., Chen, Y., Guo, M., Liu, Y., Gali, N.K., Sun, L., et al.: Aerodynamic analysis of SARS-CoV-2 in two Wuhan hospitals. Nature **582**(7813), 557–560 (2020)
9. Yao, M., Zhang, L., Ma, J., Zhou, L.: On airborne transmission and control of SARS-Cov-2. Sci. Total Environ. 139178 (2020)
10. Garaga, R., Sahu, S.K., Kota, S.H.: A review of air quality modeling studies in India: local and regional scale. Current Pollut. Rep. **4**(2), 59–73 (2018)
11. Guo, H., Kota, S.H., Sahu, S.K., Zhang, H.: Contributions of local and regional sources to PM2. 5 and its health effects in north India. Atmosp. Environ. **214**, 116867 (2019)

12. Sharma, S., Zhang, M., Gao, J., Zhang, H., Kota, S.H.: Effect of restricted emissions during COVID-19 on air quality in India. Sci. Total Environ. **728**, 138878 (2020)
13. Sharma, S., Kota, S.H.: Impacts of COVID-19 on Air Pollution. In: Integrated Risk of Pandemic: Covid-19 Impacts, Resilience and Recommendations, pp. 217–229. Springer, Singapore (2020)
14. MoEFC.: Ministry of Environmenta, Forest and Climate Change. In: Sundaray, S.N.K., Bharadwaj, D.S.R., (eds.), National Clean Air Programme New Delhi (2019)
15. Li, J., Liao, H., Hu, J., Li, N.: Severe particulate pollution days in China during 2013–2018 and the associated typical weather patterns in Beijing-Tianjin-Hebei and the Yangtze River Delta regions. Environ. Pollut. **248**, 74–81 (2019)
16. Wang, P., Chen, K., Zhu, S., Wang, P., Zhang, H.: Severe air pollution events not avoided by reduced anthropogenic activities during COVID-19 outbreak. Resour. Conserv. Recycl. **158**, (2020)
17. Zhang, H., Hu, J., Kleeman, M., Ying, Q.: Source apportionment of sulfate and nitrate particulate matter in the Eastern United States and effectiveness of emission control programs. Sci. Total Environ. **490**, 171–181 (2014)
18. Chen, G., Wang, Q., Fan, Y., Han, Y., Wang, Y., Urch, B., Silverman, F., et al.: Improved method for the optical analysis of particulate black carbon (BC) using smartphones. Atmosp. Environ. **224**, 117291 (2020)
19. Luo, Z., Zhang, L., Li, G., Du, W., Chen, Y., Cheng, H., Tao, S., Shen, G.: Evaluating co-emissions into indoor and outdoor air of EC, OC, and BC from in-home biomass burning. Atmos. Res. **248**, (2021)
20. Mousavi, A., Sowlat, M.H., Lovett, C., Rauber, M., Szidat, S., Boffi, R., Borgini, A., De Marco, C., Ruprecht, A.A., Sioutas, C.: Source apportionment of black carbon (BC) from fossil fuel and biomass burning in metropolitan Milan, Italy. Atmosp. Environ. **203**, 252–261 (2019)
21. Liu, C., Shi, S., Weschler, C., Zhao, B., Zhang, Y.: Analysis of the dynamic interaction between SVOCs and airborne particles. Aerosol Sci. Technol. **47**(2), 125–136 (2013)
22. Wensing, M., Schripp, T., Uhde, E., Salthammer, T.: Ultra-fine particles release from hard-copy devices: sources, real-room measurements and efficiency of filter accessories. Sci. Total Environ. **407**(1), 418–427 (2008)
23. Ferrer, I., Zweigenbaum, J.A., Thurman, E.M.: Analysis of 70 environmental protection agency priority pharmaceuticals in water by EPA Method 1694. J. Chromatog. A **1217**(36), 5674 5686 (2010)
24. Judson, R.S., Martin, M.T., Egeghy, P., Gangwal, S., Reif, D.M., Kothiya, P., Wolf, M., et al.: Aggregating data for computational toxicology applications: the US Environmental Protection Agency (EPA) aggregated computational toxicology resource (ACToR) system. Inter. J. Molecul. Sci. **13**(2), 1805–1831 (2012)
25. Safoutin, M.J., McDonald, J., Ellies, B.: Predicting the future manufacturing cost of batteries for plug-in vehicles for the US Environmental Protection Agency (EPA) 2017–2025 light-duty greenhouse gas standards. World Electric Veh. J. **9**(3), 42 (2018)
26. Melnick, R.: Shep. Regulation and the courts: The case of the Clean Air Act. Brookings Institution Press (2010)
27. Duvall, R.M., Hagler, G.S.W., Clements, A.L., Benedict, K., Barkjohn, K., Kilaru, V., Hanley, T., et al.: Deliberating Performance Targets: Follow-on workshop discussing PM10, NO2, CO, and SO2 air sensor targets. Atmos. Environ. **246**, (2021)
28. Becker, R.A.: Air pollution abatement costs under the Clean Air Act: evidence from the PACE survey. J. Environ. Econ. Manag. **50**(1), 144–169 (2005)
29. Lin, C.-A., Chen, Y.-C., Liu, C.-Y., Chen, W.-T., Seinfeld, J.H., Chou, C.C.K.: Satellite-derived correlation of SO2, NO2, and aerosol optical depth with meteorological conditions over East Asia from 2005 to 2015. Remote Sens. **11**(15), 1738 (2019)
30. Comunian, S., Dongo, D., Milani, C., Palestini, P.: Air pollution and Covid-19: the role of particulate matter in the spread and increase of Covid-19's morbidity and mortality. Inter. J. Environ. Res. Public Health **17**(12), 4487 (2020)
31. Rohrer, M., Flahault, A., Stoffel, M.: Peaks of fine particulate matter may modulate the spreading and virulence of COVID-19. Earth Syst. Environ. 1–8 (2020)

32. Setti, L., Passarini, F., De Gennaro, G., Barbieri, P., Perrone, M.G., Borelli, M., Palmisani, J., et al.: SARS-Cov-2RNA found on particulate matter of bergamo in northern italy: first evidence. Environ. Res. 109754 (2020)
33. Srivastava, A.: COVID-19 and air pollution and meteorology-an intricate relationship: a review. Chemosphere 128297 (2020)
34. Kanniah, K.D., Zaman, N.A.F.K., Kaskaoutis, D.G., Latif, M.T.: COVID-19's impact on the atmospheric environment in the Southeast Asia region. Sci. Total Environ. **736**, 139658 (2020)
35. Coker, E.S., Cavalli, L., Fabrizi, E., Guastella, G., Enrico, L., Laura P.M., Pontarollo, N., Rizzati, M., Varacca, A., Vergalli, S.: The effects of air pollution on COVID-19 related mortality in northern Italy. Environ. Res. Econ. **76**(4), 611–634 (2020)
36. Alqasemi, A., Hereher, M., Kaplan, G., Al-Quraishi, A.M., Saibi, H.: Impact of COVID-19 lockdown upon the air quality and surface urban heat island intensity over the United Arab Emirates. Sci. Total Environ. **767**, 144330 (2021)
37. Shokr, M., El-Tahan, M., Ibrahim, A., Steiner, A., Gad, N.: Long-term, highresolution survey of atmospheric aerosols over Egypt with NASA's MODIS data. Remote Sens. **9**, 1–23 (2017)
38. Zakey, A.S., Abdel-Wahab, M.M., Pettersson, J.B.C., Gatari, M.J., Hallquist, M.: Seasonal and spatial variation of atmospheric particulate matter in a developing megacity, the Greater Cairo, Egypt. Atmosfera **21**, 171–189 (2008)
39. Abou El-Magd, I., Zanaty, N., Ali, E.M., Irie, H., Abdelkader, A.I.: Investigation of aerosol climatology, optical characteristics and variability over Egypt based on satellite observations and in-situ measurements. Atmosphere **11**, 714 (2020). https://doi.org/10.3390/atmos11070714
40. Abou El-Magd, I., Ismail, A., Zanaty, N.: Spatial variability of urban heat Islands in Cairo City, Egypt using time series of Landsat satellite images. Remote Sens. GIS **5**, 1618–1638 (2016)
41. Abou El-Magd, I., Zanaty, N.: Impacts of short-term lockdown during COVID-19 on air quality in Egypt. Egyp. J. Rem. Sens. Space Sci. (2020). https://doi.org/10.1016/j.ejrs.2020.10.003
42. Cole, M.A., Elliott, R.J.R., Liu, B.: The impact of the Wuhan Covid-19 lockdown on air pollution and health: a machine learning and augmented synthetic control approach. Environ. Res. Econ. **76**(4), 553–580 (2020)
43. Elzeki, O.M., Abd Elfattah, M., Salem, H., Hassanien, A.E., Shams, M.: A novel perceptual two layer image fusion using deep learning for imbalanced COVID-19 dataset. PeerJ Comput. Sci. **7**, (2021). https://doi.org/10.7717/peerj-cs.364
44. Elzeki, O.M., Shams, M., Sarhan, S., Abd Elfattah, M., Hassanien, A.E.: COVID-19: a new deep learning computer-aided model for classification. PeerJ Comput. Sci. **7**, (2021). https://doi.org/10.7717/peerj-cs.358
45. Shi, F., Wang, J., Shi, J., Wu, Z., Wang, Q., Tang, Z., He, K., Shi, Y., Shen, D.: Review of artificial intelligence techniques in imaging data acquisition, segmentation and diagnosis for covid-19. IEEE Rev. Biomed. Eng. (2020)
46. Shams, M.Y., Elzeki, O.M., Abd Elfattah, M., Medhat, T., Hassanien, A.E.: Why Are Generative Adversarial Networks Vital for Deep Neural Networks? A Case Study on COVID-19 Chest X-Ray Images. In: Big Data Analytics and Artificial Intelligence Against COVID-19: Innovation Vision and Approach, pp. 147–162. Springer, Cham (2020)